D1469694

ROBOTICS

ANNE CARDOZA AND SUZEE J. VLK

TAB BOOKS Inc.

Blue Ridge Summit, PA 17214

SIDNEY B. COULTER LIBRARY
Onondaga Community College
Syracuse, New York 13215

FIRST EDITION

FIRST PRINTING

Copyright © 1985 by TAB BOOKS Inc.
Printed in the United States of America

Reproduction or publication of the content in any manner, without express
permission of the publisher, is prohibited. No liability is assumed with respect to
the use of the information herein.

Library of Congress Cataloging in Publication Data

Cardoza, Anne.
Robotics.

Bibliography: p.
Includes index.
1. Robotics. I. Vlk, Suzee. II. Title.
TJ211.C26   1985   629.8′92   85-4651
ISBN 0-8306-0858-3
ISBN 0-8306-1858-9 (pbk.)

Front cover photograph courtesy of Automatix, The Robotic Systems Co.

Contents

Introduction

Some of the smartest and wealthiest people in the world are employed in the robotics industry. As a result, this book was written to encourage the reader to examine the inroads that robots are making in the quality of life. Robots have introduced "technoshock" into the workplace. This term defines the conspiracy between computers and robots to move industry at a much faster pace. Industry now requires labor to become fast-track and high-speed (computer-paced) in order to stay affordable, profitable, and consistent in quality.

The robotics industry has been around for over 20 years. Today it is impacting the machine tool industry in the same way as sewing machines did the garment industry at the turn of the century. Robots will always remain machines that carry out people's ideas at a speed and uniformity that no human can imitate.

Terms like *artificial intelligence, machine vision,* and *battlefield robotics* were coined by people who read science fiction and play video games. The video-game mentality has been transferred to battlefield, industrial, and personal robotics. Most mobile robots are operated by remote control with joysticks. From such seemingly oxymoronic unions of play and work, great advances in robotic technology have occurred. Robots help mobilize the handicapped and guard prisons. They take the place of human soldiers, handle food, and even design fashions.

Robots have not produced any real dissatisfactions among end users. The best of robotics companies have attracted important venture capitalists, putting impressive support behind individual companies. The military started producing robots; soon robot manufacturers were leaping into battlefield and security robotics.

Theories on the overall population of robots in the world are as diverse as the variety of automated equipment found in factories. In England, *Production Engineer* magazine predicted that applications for robots would increase more than 50 percent per year. Currently 45 percent of installed robots are in Japan, 35 percent in the United States, and the rest divided mainly among the European countries.

Financial analysts and brokerage houses have predicted the growth of the robotics industry as explosive, with many shakeouts. In 1979 *Fortune* magazine projected a $700 million to $2 billion robotics industry by the end of this decade. Other brokerage houses ran their predictions up to $4.6 billion by 1990. As robotics technology advances, improved methods for speeding up tasks will be seen. Future developments for robotics technology are in vision, tactile sensing, hand-to-hand coordination, human/robot voice communications, mobility, self-diagnostics, and energy-conserving design. Robots are being created that can understand and follow simple, spoken commands. The field of artificial intelligence has joined with machine vision to create the new technology of computer-integrated manufacturing.

The new industrial revolution is one of total factory automation run by computer/robot cooperation. Human factors research in robotics has made real the science fiction profession of "robopsychologist," an invention of science fiction novelist, Isaac Asimov, in his novel, *The Space Merchants*. The term described a fictionalized occupation invented by science fiction writer, Frederick Pohl.

Science fiction occupations such as "industrial androidologist," "robotics investigative reporter," and "machine vision and artificial intelligence analyst" today are realities or real possibilities for careers. The idea of such jobs being created and the reality of robots in the factory and the home has resulted in a profound sense of responsibility that has evolved within labor.

Each segment of industry has fully accepted the awakening of responsibility to people as robotics expands. This sense of responsibility is a response to technoshock. The importance of safety, human factors research, and the quality of human life has taken on new dimensions. The factory integration of robots and computer-aided manufacturing systems have created opportunities in human factors engineering. The results are jobs for specialists in training, human performance analysis, safety, and media.

Robotic factory integration is concerned with ways of designing machines, operations, and work environments so that they match human capabilities and limitations. Human factors engineering is a new area to the manufacturing industry. It exists only because robots have been invented. Technoshock results when robots evolve faster than humans can adapt to the change. It is the human who must take an outdated skill and bring it up to the level where it is acceptable to technology. Then this same person must teach his updated skills to a robot who will produce faster, cheaper, and better than the human worker.

The introduction of one robot into a factory affects departments all over that factory. People have to be hired to manage the human factors issues concerned with the physical and emotional work environment, worker-machine interface, job design, selection and training, maintainability, safety, management, and communications. Because one robot was installed, project planners have to be hired. Job designers are recruited to create smooth workflow.

Someone has to teach the robot the skills of a welder, painter, loader, inspector, or electronics parts assembler. Robots are no longer destined for the factory floor. They are now mobile and out on the battlefield and in space.

Program evaluators must coordinate the human factors involved in factory integration. All robots are human-machine systems because they are repaired or operated by people. Robots change the way managers work by changing organizational design.

New management-union relations are created when a robot is installed. Communication requirements change. Technical writers are hired to write instructional manuals for robot operators, repairers, and applications engineers. Producers are hired to create industrial films for employees. Robots attract independent consultants and freelancers.

Occupational safety technicians are called in to inspect automated factories and make sure material handling devices have warning systems and other safety measures for operators. The key human factors issues are in training, maintenance, and safety.

Another motivation for writing this book is to show that the robot is the most advanced tool used to speed the progress of humanity. Machines have been invented to take the bite and burden out of work. The human brain will always be paid to design and try out new ways to do everything.

Ideas are worth a fortune. Ideas that machines can produce inexpensively and faster are worth even more. Humans who adapt to technoshock will find that understanding robotics is a vital strategy for success in the world of high technology. In robotics, there are no fixed or conventional methods, no recipes. Progress comes from skillfully planning and directing operations.

Robotics research and development is a market-motivated technology, the new wealth of nations. Most of the predictions about robots by industry have been realized.

A 1977 meeting of the Robot Institute of America resulted in a Delphi forecast. Reported by the "father of robotics," Joseph F. Engelberger, President of Unimation, Inc., there are a number of desirable attributes with which robots should be equipped.

Sensor-controlled movements of robots such as vision and tactile sensing are desirable traits in robots. Robot users felt that a cost of $7000 could be justified for simple vision and $2000 for touch sensing in a robot. The participants agreed that simple vision rated the highest priority for research and development in robotics. All sensory capabilities were seen to be available before 1985. A shift was seen into the middle and late 1980s to more sophisticated computer control. With this sophistication, robots would be capable of coordinate transformation and sensory feedback control. The market for industrial robots in 1984 was predicted to be about $200 million with a 25 percent growth rate per year during the 1980s. Engelberger predicted a market of approximately $3 billion for the combination of Western Europe and the United States in 1990.

The Delphi forecast also listed 20 robot attributes that were predicted to become a reality by 1984. They have all come true. These are the attributes:

1) Work space command with six infinitely-controllable articulations between the robot base and a hand extremity.
2) Fast, hands-on instinctive programming.
3) Local and library memory of any size desired.
4) Random program selection by external stimuli.
5) Positioning repeatability to 0.3mm.
6) Weight handling capability to 150 kilos.
7) Intermixed point-to-point and path-following control.

8) Synchronization with moving targets.
 9) Compatible computer interface.
10) High reliability.
11) Rudimentary vision via orientation data or recognition data.
12) Tactile sensing via orientation data, physical interaction data, or recognition data.
13) Multiple-appendage hand-hand coordination.
14) Computer-directed appendage trajectories.
15) Mobility.
16) Minimized spatial intrusion.
17) Energy conserving musculature.
18) General purpose hands.
19) Man-robot voice communication.
20) Inherent safety.

It is remarkable that between 1977 and the present these twenty attributes have been realized. For example, Unimation's PUMA robot has a vision system that permits the robot to assemble parts. With "eyes" to see objects, the robot can locate, identify, and lift randomly strewn parts. This required the invention of automatic feeding and orientation components. Almost every one of these robot attributes that may not have existed before the 1970s have created new jobs for humans. Engelberger predicts the big activity for robots will be in assembly during the middle of the decade.

Currently more than half of the workers in the United States are engaged in various forms of information or knowledge processing. Knowledge is a more salable commodity than energy or food. The robot is the manufacturer's tool, much as the computer is the knowledge worker's tool. The ascendancy of robotics has occurred because the computer is an amplifier of the "smart" robot. The robot, in turn, is an extension of the brain, limbs, and senses of the human worker.

Artificial intelligence has given robots the abilities to process knowledge, not just information. This fact alone has created the era of the "smart" robot. Artificial intelligence is creating value in robots by transferring the brainpower of the knowledge workers to machines that think logically. Robots use the same movements at the same speed as human beings to perform a job. Computer or radio-powered robots don't consume raw materials. They use far less energy than electromechanical or hydraulic-powered machines.

This book was written to alert the general reader to the revolution of the "smart" robot. You can get in on the ground floor of a new technology that has witnessed explosive growth. This expanding, worldwide $157 billion market for innovative software programs for robots and for technology and service has provided the incentive for industry to speed up developments in robotics. Factory automation affects everyone.

Efforts need focus. Each person needs his or her own plan of action of how to incorporate the world of robotics into his or her lifestyle. Technical knowledge must be pooled and available to everyone. The current state of robotics is analyzed in this overview. Many of the major robots on today's market are discussed.

The uses of robots are detailed, and the opportunities for the newcomer in robotics technology are discussed. The section on education and training shows you who offers training, the cost in terms of time and money, and how to begin to apply that instruction. Sections at the back of the book include a glossary of robotics terminology and sources of additional information.

There are jobs in robotics for people from all kinds of backgrounds—linguistics, programming, technical writing, electronics technology and testing, repair services, sales, management, engineering, drafting, international trade, training, and recruiting. To the reader looking over the field, working with computer-controlled machines means teaching machines to perform jobs.

Knowledge itself is an international bartering tool. Robotics, artificial intelligence, and digital computers are all part of the tools humans have to pool and use at our disposal. Robots aid humans in our efforts to span time and space, reaching out to make the most efficient use of each.

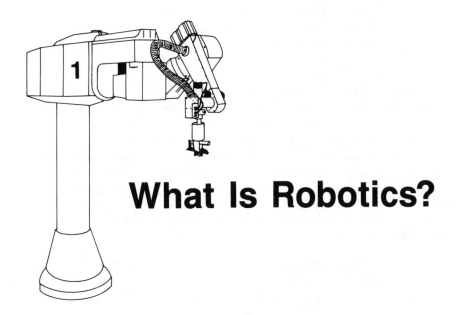

What Is Robotics?

A robot wheelchair built by a team of students at Drexel University in Philadelphia talks back to a disabled person, confirms instructions, and announces the progress of a wheelchair excursion. It is controlled by puffs of breath and/or the slightest voice command. The robot memorizes house plans. It takes the person anywhere in his home without banging into walls or tumbling down stairs. The robot acts as the person's eyes and limbs.

Robots are machine tools that may be found in the dirtiest of industrial work sites, where they perform spot welding and materials handling. The most frequent industrial applications for robots are spray painting and performing assembly applications. Personal robots are used as security devices in homes. Promotional robots walk shopping malls, displaying advertisements. Robots guard prisons. In Japan, robots build most of the prefabricated houses and perform heavy carpentry applications, working 24 hours a day. Robots are used by the military and police departments to defuse live bombs and stand sentry duty.

The word *robot* comes from the Czechoslovakian word *robotnik*, meaning "slave." It was coined by Czech playwright Karel Capek, on the advice of his brother, for his science fiction play, *R.U.R.* (Rossum's Universal Robots).

Years later, the word *robotics* was coined by Isaac Asimov in one of his novels about the manufacture of androids. Thirty-five years later, the term *robotics* was applied to an entire industry and programs of higher education. Isaac Asimov worried about the prospect of extending the human mind into robots to give machine tools human-level intelligence and even consciousness. Asimov's novels dealt with unfriendly machine intelligence.

To prevent robots from hurting people, Asimov proposed guidelines in his novels for robot programming. He wrote his renowned "Three Laws of Robotics." Asimov's laws prevent a robot from harming a person or, through neglect, allowing disaster to come to a human being. Asimov recommended that similar laws be built into artificial intelligence machines to prevent them from being "insubordinate." Computers have been killing off humans to prevent their own disconnection in numerous science fiction movies from 1927 to the present.

Frederick Pohl, another science fiction writer, created the profession of "industrial anthropologist" in his book, *The Space Merchants*. Pohl applied the term to the anthropology of industrial space colonization. Today the term *industrial androidologist* has come to be applied to a social scientist who studies the impact of robots on industrial society, as well as the impact of artificial intelligence on robots.

Joseph Engelberger, world-renowned robot inventor and entrepreneur, was an avid reader of Isaac Asimov's novels. His science-fiction approach to robotics proved to be the motivation that sent him on a lifelong quest that a eventually earned him the title "The Father of Robotics." When the Robot Institute of America (RIA) decided that there must be an agreed-upon definition of robotics, they consulted with Engelberger and came up with the following definition:

> A robot is a reprogrammable multifunctional manipulator designed to move material, parts, tools, or specialized devices through variable programmed motions for the performance of a variety of tasks.

The RIA's definition of robotics has been accepted worldwide. Technological impact on workers is the most important problem for the robotics industry to solve today.

WHO'S BUYING THE ROBOTS?

When robots are installed in a factory, the workers are usually told they have better things to do than be a machine tool. Workers then are retrained, at their employer's expense, in robot maintenance or programming. This act is repeated in factories around the nation at a slow but steadily increasing rate. Companies with robots consist of the robot builders and the robot users. More jobs exist with the users of robots than with the builders. A third type of company, the robot component manufacturer, is steadily gaining ground. Such firms produce vision systems and other sensory parts for robot manufacturers.

United States buyers purchased about 7000 personal robots in 1983, a market worth $14 million to $18 million to retailers, according to R.B. Robot Corporation, a firm in Golden, Colorado, that conducted the research. However, the largest market for robots is in industry. During the past decade, labor costs in this country have grown by approximately 250 percent, but the cost of making robots has increased by only 50 percent. By 1990, robots will create 32,000 to 64,000 jobs. There will be between 50,000 and 1 million robots on the job automating factories, according to robotics author/researcher H. Allen Hunt. Hunt compiled his statistics primarily in Michigan, where currently over 1000 students are enrolled in robotics technology training. In his 1983 book, *Human Resource Implications of Robotics*, Hunt predicted that there will be two jobs for every three new robots, and each robot will eliminate two jobs. The Japanese estimate that six jobs will be terminated by each robot.

In the automobile industry today, robots can replace up to six workers if the robots are operated around the clock, as they are in Japan. Quality control increases fourfold with robots. In 1983 there were 33 percent fewer automobile workers than when the decade started. Jobs for teachers of robotics technology are predicted to increase. By 1990, up to 20 million Americans will have to be retrained entirely or find their job skills obsolete, according to Pierre S. du Pont IV, former governor of Delaware and chairman of the National Task Force on Education for Economic Growth.

By 1987, the number of electronics workers will grow by 115 percent, according to the American Electronics Association (AEA). All this is possible because of the most advanced computers, also known as "artificial intelligence" machines. Because the machines are capable of planning operations, producing other robots, making intelligent decisions, and communicating with other machine tools—and because the price of robots is likely to drop—robotics will be the accepted technology in factory automation.

HOW THE CAD/CAM INDUSTRY SUPPORTS ROBOTICS

A spin-off of the robotics industry has created a new technical art form, *computer-aided design* (CAD), which is automated drafting. Another concept is *computer-aided manufacturing* (CAM). These two often are grouped together as the CAD/CAM industry. Graphic design is performed with the assistance of a computer for computer-aided design, and computers are hooked up to control robots in computer-aided manufacturing.

In the CAD/CAM industry, the notion of "automate or perish" is the rule. Picture and image processing are considered almost as important as language processing. Computer-aided manufacturing utilizes artificial intelligence to analyze every industry, from satellites to medical imaging. Computer-aided design is used in medicine for diagnosing illness; computer-aided manufacturing is used to automate a factory.

Research in CAD/CAM as applied to robotics is taking place in three phases. An experimental phase tackles hardware architecture, so that robots are able to distinguish the boundaries of objects. Machines need vision. Computer-aided design utilizes display generators and works from an image database.

A pilot model is designed in the second phase; a final phase joins computer-aided manufacturing with the "fifth-generation" computer—artificial intelligence. The obvious application of artificial intelligence is the goal of the robotics industry. Artificial intelligence creates a robot that can see, understand, and act under new circumstances.

The bulk of robotics research and development in Japan is set up in a Robotics National Project; in the United States, each robotics manufacturing company is working on its own. No national project currently exists, but experts predict that there will be a national project by the end of the decade. At present the only means of networking in the robotics industry is a few associations, a few conferences, the robotics educational programs, and a handful of fiercely competing robotics trade publications.

A robotics industry "conspiracy," that is, the effort to network, is only beginning to be aired. Eventually, the image understanding system of CAD/CAM is expected to utilize artificial intelligence to store about 100,000 images. In this, as in voice recognition, the Japanese are in the lead in research with their own National Robotics Project.

A decade ago, the Japanese announced a national project on *pattern information processing systems* (PIPS) research. In Japan, 25 years of research in artificial intelligence has been focused on developing applications for robotics. In the United States the concept of the "fifth-generation" computer is being used to improve the sensory capacities of industrial robots through new developments in the CAD/CAM industry. Machine vision and speech recognition systems for robots are produced by the CAD/CAM industry.

No matter how advanced robotics is to become, a robot is always going to be a programmable machine. Some industrial robots are powered by complex computers, but they still perform a limited sequence of motions. Robots replace human labor. People must watch a robot indirectly while it is running. Each job that in some way involves overseeing a robot through every step of its operation requires humans with varying levels of skills. For every job involved with programming, designing, or repairing robots, a program operates to train persons who will train other persons to program robots and the computers that control the robot's actions. Software must be updated. There is still a chain of labor.

THE LIMITS OF ROBOTS: OVERCOMING UNFAMILIARITY

Robots are placed in regimented environments to perform simple, repetitive, unskilled tasks at the present time. Unhooked from their computers, the artificial intelligence software removed, robots become unthinking, untiring machines that have a 40,000-hour (20-year) worklife. Robots must be overhauled at intervals. Computer repair is a multibillion dollar industry, and robot repair is quickly catching up as more robots are installed. Robots may work in conjunction with other computer-controlled programmed machines that are synchronized by their control systems.

In this last case it is often not possible to back up these machines with human operators. When one machine stops, all other machines will cease to operate. *Down time* is a function of the mean time between equipment failures and the mean time needed to repair the failed robot. Industrial robot applications can be costly unless expected down time remains under 2 percent.

Robots are handled very differently from human labor. The cost of a robot is capitalized and depreciated over its useful 20-year lifespan. The depreciation shows up as a cost on the company's profit-and-loss statement. Robot investment is risky. Massive investment in robotics for any given year can be premature, because new robots which can do the work better and more cheaply enter the market each season. As a direct result of an expensive robot investment that becomes obsolete too quickly, blue-collar workers are laid off. When human workers want to know why they are the first to be laid off when a robotics investment proves disastrous, they are usually told that the business borrowed money to buy a robot and incurred a long-term liability. Now that liability must be repaid.

The use of robots in business has been justified because industry is decentralizing into tiny, highly technical, easily controlled units of manufacture. Robots managed by numerically controlled computers that perform various kinds of machining and processing reduce set-up time. As a result, lot quantities of a manufactured product can now be smaller than they were in the past. As quantities of a product grow smaller, fewer human workers are needed. Robots are used to target inventory cost as a needless, unproductive expense. Every stroke of a robot arm per minute (or any other unit of time) can be recorded and measured. With inventory stocks low, incoming quality is taking a primary position in industry.

The most urgent need today in robotics is advancement in robot vision and speech recognition. The robot must "see" and understand speech in order to pick up the right component and process a product. Otherwise robots must be continually teamed with human operators.

The goals of the robotics industry are to free operators for judgment tasks and to let the robots do the menial or repetitive work. At present, robots cannot be inspectors because they lack the right vision systems. Robots are not able to make judgments about the quality of a product. No robot can sniff a can of food and judge whether the substance is fresh.

Robots are a new technology. Technical people are still unfamiliar with the appropriate selection criteria, operating parameters, and maintenance requirements. Four levels of robotic development exist to serve the needs of industry and the personal robot market. They may be summarized as follows:

1) Personal computer robots. These machines have no sensory capability. They are used as home security devices, promotional robots, or as toy house servants.
2) Programmable robots with some sensory capability. One example would be the Cincinnati Milacron CONCITE, which has a simple optical sensor, or robots with pressure sensors that say touch or no-touch.
3) Sensory robots that have manipulative ability, such as the two-armed General Electric PRAGMA A-3000. Sensory robots include robots with high-resolution touch sensors that can feel variations in shape and distinguish shadows from real objects by machine vision systems.
4) Artificial Intelligence. These "fifth-generation" computers are attached or built into robots which have sensors, manipulators, and "intelligent" behavior. Machines such as Jet Propulsion Laboratory's Mars Rover prototype, which traveled hundreds of miles across the Martian terrain without human interference, integrate artificial intelligence and robotics. These robots will be used in space mining and space materials-handling jobs in the next two decades.

ADOPTING STANDARD ROBOT LANGUAGES

All of these robots interact with computers or are driven by radios. These machine tools use a variety of robot languages, which include AL, AML, Anorad, Autopass, Emily, Funky, Help, Maple, MPL, PAL, RCL, RPL, Sigla, T3, and VAL. (The IBM Advanced Robotic System shown in Fig. 1-1 uses the IBM-developed robotics programming language AML.) There is a very high demand for new robot languages and new software to control robots.

The particular robot language determines in large part how well the robot performs. The sole purpose of a robot language is to make it simple to define a job that a robot has to do. Robots are programmed off-line by computer-type languages. *Off-line* refers to programming without requiring that the actual robot and work pieces be in front of the programmer. A programmer operating an off-line programming systems must be able to visualize accurately just what will happen when his or her program is being executed.

5

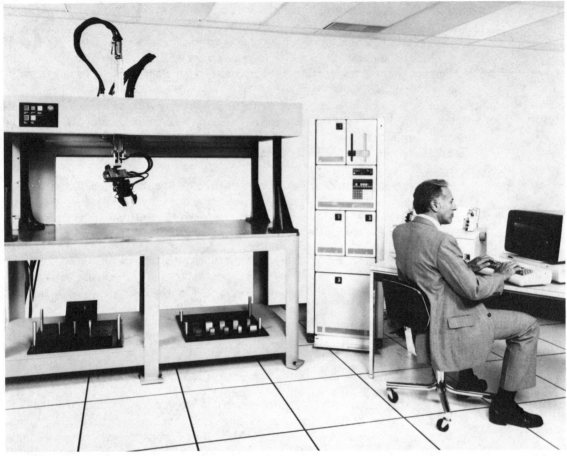

Fig, 1-1. Controlled by an IBM Series/1 computer and programmed in AML (A Manufacturing Language), a powerful programming language developed by IBM specifically for robotic applications, the IBM 7565 Manufacturing System can be used in industries such as electronics, automotive, aerospace, and appliance fabrication. An operator can program the system either through the keyboard/display or with the hand-held teach pendant shown at the top left corner of the manipulator frame. Courtesy International Business Machines Corporation.

Robot language development is a worldwide industry that is predicted to reach the multi-billion mark just as computer language development has in the last decade. People craft products and each piece is unique; robots produce every component the same unless a human programs them to make each different from the next. What a robot does depends on the robot language. To production engineers trying to get robots to do anything a person can, but faster and with less variation (when variation is seen as error), the trend is toward adoption of standard languages. Off-line programming systems are still in the Stone Age of research.

In the United States robot language research has been performed in the Language Standards Committee of ICAM, the U.S. Air Force Integrated Computer-Aided Manufacture program, dedicated to aerospace production. The large robot manufacturers such as Unimation, Cincinnati Milacron, and IBM are also engaged in language development.

Giant-sized users of robots, such as General Motors, are trying to find better solutions to programming problems with large populations of off-line robots. The company's robot suppliers come from different firms. Massachusetts Institute of Technology (MIT) and Stanford University have always been the world leaders in the academic side of off-line programming.

In Europe, Italian robot manufacturers provided manipulator languages. Britain has for years engaged in robot language research; similar work is underway in France and Germany. The Computer Integrated Manufacture (CIMP project of ESPRIT) and the EEC Standards Working Group are involved in robot language research. A Japanese group within a robotics interest group, Computer Aided Manufacturing International (CAM-I) is working on robot languages at the present time.

One of the biggest problems to solve in the robotics industry is how to provide links to computer-aided design databases. Another giant-sized problem to solve is how robots can handle small batch production. Robot languages are classified according to increasing power or level of abstraction. Languages are needed to solve these problems by creating and maintaining a computer representation of the robot's environment.

The robot must find a way to describe three-dimensional objects while carrying out its job. The robot needs to find ways to make inferences about relationships between objects. In short, the language that gives a robot geometrical reasoning will make a profit. A robot needs to make use of past knowledge about the parts it sees in the present. This is called *model-driven pattern recognition*.

Then the robot has to define the correct response to sensors in order to provide information about the position and place of the work pieces. Robotics is all about the way human and machine interrelate, an action which usually is called *interfacing*. Languages have to be tailored to specific needs. Parts have to be standardized. The outlook for robots lies in the development of better software.

Reliability will make or break the robotics industry. In order to examine the trend of better, cheaper, and faster robots coming into the market each season, the impact of robotics on people needs to be examined in a historical light. Although the word *robotics* is a 20th century coinage, the idea of a surrogate to do the work is as old as prehistory.

From the time Swiss craftsmen of the 1700s created robots with music box innards which drew pictures and wrote letters, the idea of cheap labor has remained with us. In the following section, the history of robotics unfolds as the portrait of an industry that has created machine tools to fill a need which humans can no longer fill.

BEGINNINGS OF 20TH CENTURY ROBOTICS

Serious cocktail parties promote knowledge fusion among people with similar work interests. It was a serious cocktail party one evening in 1956 when George C. Devol sat with Joseph Engelberger, talking about Isaac Asimov's conception of robotics.

Devol, an inventor with patents in the electronics field, too numerous to count was anticipating a patent for his "Programmed Transfer Article," as some robots were called in the 1950s. Engelberger, the self-made entrepreneur, was an avid reader of science fiction novels about all-consuming robots, primary those of Isaac Asimov.

The two men were introduced, and from this chance cocktail-party small talk Devol and Engelberger created big ideas and nurtured those ideas through every step of their development, overseeing their projects like parents watching the growth of a child. Each man went home to create a new age—the generation of robots. Engelberger went on to earn the nickname "The Father of Robotics," and Devol's U.S. Patent 2,988,237 was granted in 1961. This patent heralded the android birth of the first Unimate robot, directly inspired by the humanoid machine creations of Isaac Asimov's science fiction. Unimate was no fiction.

In the same year of that cocktail party, 1956, a formal logic machine was unveiled at Dartmouth College. Programmers were trying to make artificial intelligence systems play chess using formal logic. A program called Logic Theorist, designed to prove logical propositions was presented at a conference on artificial intelligence. In a contest against famous mathematicians such as Alfred North Whitehead and Bertrand Russell, the artificial intelligence machine proved itself superior in mathematics by coming up with a proof missed by these two human geniuses.

The machine was unpredictable, however. Programmers could not always know how the program would perform under variable conditions. It gave a mathematical proof for a Whitehead-Russell theorem—despite the fact that the machine was not asked to look for a new proof the two mathematicians had missed.

The path that the Logic Theorist machine would follow was to be linked with "smart" robots. Isaac Asimov's novels of humans extending their limbs and brains with machines had come true. The "thinking" robot was no longer science fiction. The idea caught on with George C. Devol. In the 1950s Devol was struck by the skyrocketing rate of automation obsolescence and by the multitudes of blue-collar labor needed for repetitive assembly line jobs. Each year roughly 60 percent of the budge of most industries was used for payrolls. As comparisons sifted through Devol's mind, he wanted more than anything to see universal automation perform diversified jobs and be protected against model-change obsolescence.

During World War II he organized a company to produce radar countermeasure systems and specific products for the OSS, the forerunner of the CIA. By 1950, Devol had patented a new form of magnetics recording and had worked on the development of large-scale random access memories and ultra-high-speed printers. He worked night and day to apply digital control to a "manipulator" which could be programmed on the job. The machine tool was *multi-axis*, meaning it could access components from many angles.

As 1954 passed, Devol was deeply concerned with over 40 patents on a tooling and transfer system, which would later be dropped in favor of a robot. Devol never did acquire the very first robot patent. A British inventor named Cyril Walter Kenward held a 1957 British patent, number 781,465. Unimation and AMF, another United States robot manufacturing firm, gave Kenward a cash settlement. Unimates were built by hand in 1961. The first robot was sold to the Ford Motor Company, to be used for overseeing die-casting machines.

This first U.S. robot was sent to a company that didn't call it a robot. In the 1960s industrial robots were called "univeral transfer devices." (Ford still calls its robots UTDs.) The first robot is on display at the Smithsonian Institution.

Unimate had a rival robot, Versatran, made by AMF. These universal transfer devices were used as human labor surrogates to handle the dirtiest, most noxious jobs involving repetitive movements. The robots poured hot liquid chemicals into molds, cast dies, cut metal, and used their weight as presses. The robots never manipulated tools. They handled parts or components in assembly-line formation.

Except for the rapid advances in software and on controllers, the Unimate robots are mechanically similar to the first machines manufactured in the early 1960s. These historical robots were operated hydraulically. Unimate formed the foundation upon which other robots became spin-offs of hydraulics technology. Even more numerous were the look-alikes of other robotic designs which currently are in operation in most of the world's major industries.

One such robot is ASEA's IRb. ASEA offered the first all-electric-drive IRb 6 robot in 1973, and the IRb in 1974. These robots were among the first all-electric machines. A survey published by the IFS in 1972 listed six electric robots out of a total of 133. All but one of these robots have disappeared. Only Yaskawa continues to sell today, as Motoman. Currently the ASEA robot is designed for operation in difficult environments. Complex programs can be compiled with dissimilar items easily and rapidly. The robot offers different operating sequences for items, additional steps, integration of the operating sequence, and other features unknown in the 1970s. The robot is programmed by means of a "dialogue" with the control system via a portable programming unit. Today's robot contains a control system that poses questions in plain language on a typewriter-like keyboard. The operator answers by pushing the buttons providing the correct answer. The robot can be ordered to move according to rectangular, cylindrical, or wrist-oriented rectangular coordinate systems.

ROBOT CONSULTANTS ARE IN DEMAND

Androidologists are called upon by industry more often than anthropologists. Today, nearing 73, Devol runs Devol Research Associates, a Fort Lauderdale-based consulting firm which helps robot users evaluate and implement robotics concepts.

His consulting business is growing toward becoming a base from which robot research development centers share ideas for growth. Its purpose is to combat the shortage of skilled researchers and developers of faster, smaller, and less rigid robots. Devol's research spans electromechanical muscles, voice control, environmental sensing without television cameras, and the possibility of robot rentals for small manufacturing firms.

Joseph Engelberger, founder of Unimation, Inc., was named "The Father of Robotics" in 1983 by *American Machinist* and the Robot Institute of America, for pioneering the industrial robot, creating a new industry, and providing a novel approach to industrial automation.

Engelberger was only 31 when he met Devol in 1956. At that time Devol was a free-lance inventor for Sperry Gyroscope and General Electronic Industries. Engelberger was a general manager of a small aerospace division at Manning, Maxwell and Moore, in Stamford, Connecticut. Engelberger went into experimentation with robots as a sideline for his operation. By 1957 Manning, Maxwell and Moore was getting out of the aerospace business. Engelberger enlisted the aid of Norman Shafler, the founder of the military vehicle manufacturer, Consolidated Diesel Electric Company (now Condec), to purchase the 160-worker aerospace division.

Engelberger set up his business in a garage and began to turn out aircraft components. Consolidated Controls built its first robot, originally called a "universal helper" and then named "Unimate" by Engelberger and Devol in 1959.

The first Unimate was installed in a General Motors plant in 1959. By 1961, the first production Unimate was sold. At first the robot was used to load and unload machines. The first few robots sent to General Motors and to Doehler-Jarvis worked with heated die-casting machines. Unimation, Inc., was founded in 1962 when the Pullman Corporation and Condec entered into a joint venture.

By the early 1970s, robot stories became widespread in the media. The Robot Institute of America opened its doors in 1974 to anyone interested in acquiring information about robots. It was the first trade association in the United States to discuss robots with industry and the public.

The first generation of robots in the 1950s had hard-wired controllers and either pegboard or wire memories. A decade later, the robot was recognized as a cost-effective approach to automation. The first robots had to set the stage for a flexible design, so that enormous changes could occur annually in processing, product design, and production timing.

In 1972 Unimation had a Japanese licensee, Kawasaki, which put robots into Nissan Motors. Today Japan uses more robots in industry than the United States. At this point of history—1972—Japan exploded in a robot revolution that took over virtually every major industry.

All this robot development evolved as a direct result of changes in human population during the 1960s. Until 1960, the population of the United States grew each year at an increasing rate. The turning point occurred in 1960, when the rate at which the population was growing formed a plateau and then began to decline at an increasingly sharper angle.

Between 1970 and 1980 the work force expanded rapidly. The 75 million people born after World War II entered the work force beginning about 1965. After 1970 the divorce rate grew gradually to 60 percent, and large numbers of women entered the work force. Eighty percent of these women remained in low-paying clerical and sales jobs, whereas many of the males entering demanded higher-paying technical jobs—with retraining at company expense. Affirmative action and access to untapped labor programs soon blossomed to induct women and minorities into robotics technician training programs.

At the beginning of the 1970s the Vietnam war ended, and large numbers of veterans poured into the labor force. As the decade ended, the economy was unable to find jobs for a work force that had expanded more rapidly than the number of available jobs.

Each company had to deal with its own decisions on how to expand. The choice was whether to automate with robots or to lay off workers. Along with robots came the risk of investing in a costly technology. For security reasons most businesses chose to stand by what was familiar in the past. The promise of profits from automation was not calming enough to assuage the fear of facing the unknown.

It costs at least $10,000 to install a robot in a factory, and then only if little modification of existing facilities is required by the robot. Normally such installation costs have run as high as three times the robot's purchase price.

There are considerable depreciation costs. A robot usually can't work beyond 7 years without overhaul. Many robots in use are depreciated over 10 years, but are overhauled at the end of 5 years of two-shift operation. Such overhauls can cost 25 percent as much as the original cost robot. Experts agree that robots need frequent overhauling, usually after 15,000 hours of operation. The work life of a robot is generally about 40,000 hours.

American business had to make a choice in the light of these costs.

In 1970, American business chose human labor over robots. Labor was plentiful and cheap. This use of labor instead of capital looked good superficially. Labor was adaptable. Robots were still inflexible. As the decase wore on, the gross national product expanded, and human labor stretched that growth.

Robots stagnated in the 1970s. Increasing productivity by laying off workers was a waste of time when the "baby boom" generation was fresh out of school and ready to work. All this human labor could be tapped without the technological risk and capital investment that robot installation involved. To compound the dilemma, however, human labor was demanding higher salaries and more responsibility.

In 1970, the first symposium on industrial robots in Chicago became a national event. The year 1972 was the turning point for Japan as a robotics revolution exploded in Japanese industry. Dramatically, Japan became the first nation to form a robot association, The Japan Industrial Robot Association (JIRA). Four years later, the United States opened the doors of the Robot Institute of America (RIA) in Dearborn, Michigan. In 1977, the British Robot Association was formed. Today robot associations exist in countries that are heavy users of industrial robots: Australia, Belgium, China, Denmark, France, Italy, and Sinapore.

PATTERN RECOGNITION: THE BIGGEST HURDLE

An expansion in applications occurred in robotics between 1950 and 1960. Robot vision design was started in the 1960s by artificial intelligence researchers at SRI International in California and at Edinburgh University in Scotland.

Patten recognition was the biggest hurdle to leap in research and development. Nottingham University in Britain developed in 1972 the SIRCH machine, which was able to recognize two-dimensional parts in a random pattern after programming. It took another 8 years before this machine was applied to industrial uses.

Industry had failed to understand the basic techniques necessary to be learned before a machine operator could communicate with the robot controllers that existed in the early 1970s. General Motors was the first corporation to use robot vision in industry. The company received material about SRI's developments at the General Motors Technological Center. General Motors selected SRI's Consight system and installed the robot vision machine at a foundry in St. Catherines, Ontario, Canada.

A conveyer belt moved hundreds of castings as a T3 robot stood over the "shake-out" area. The robot "saw" and "recognized" the proper casting and placed it in its proper compartment. Between 1965 and 1975 another expansion in robotics occurred. This time, vision and artificial intelligence was emphasized. Cincinnati Milacron announced its T3 robot for sale in 1975. At that time the T3 was called the 6CH robot. For the first time robots were controlled by computers instead of hydraulic or electric systems.

Between 1975 and the present, the range of computer-controlled robot applications has multiplied rapidly. By 1982, France's Renault used a complicated vision system to program a Renault V80 robot to lift crankshafts from a palletized pile and load them into a grinder.

TACTILE SENSING AND VISION SYSTEMS

Touch and force sensing by robots has outdistanced even vision systems as the foremost advance in robotics. Hitachi came out with a Hi-T-Hand robot in 1974. This machine used force feedback in the robot's hand to guide tiny pins into minute holes. Touch sensing gave the robot another sense. It already had vision, and now it had touch—tactile senses so it would "know" the pressure of its own strength on objects.

By 1976, robots were made capable of "part mating." Research was underway at the Charles Stark Draper Laboratories in Cambridge, Massachusetts; this work involved research using a passive device that gives no feedback. It is called the *Remote Center Compliance* (RCC). The RCC component gave robots tactile ability. The device allows "distortion" to be introduced into a robot's pattern. For example, as a pin is pushed toward an opening, the RCC allows the robot fingers to line the pin up laterally and rotationally with the opening, so the pin can be inserted with only a few microns clearance.

By 1983, robot hands capable of playing simple melodies and trilling notes on a piano many times faster than human hands had been built in a Japanese university. These experiments are given practical applications in devices to help the disabled.

The 1983, Computer-Aided Manufacturing International (CAM-I), a robotics interest group in Arlington, Texas, took a plunge into robotics software, building upon the work of Professor Norio Okino and CAM-I's Japanese affiliates. CAM-I's goal for current software projects is to create an information exchange between robots that includes pattern recognition and function enhancement. The special interest group tries to clarify robotic software configuration definitions and works with advanced methods of robot control, called *algorithms*.

CAM-I seeks to standardize a robot language. The first step in doing this is to clarify software requirements and functions, then prepare a detailed plan for a standard robot language. Professor Okino of Hokkaido University in Sapporo, Japan, has found 22 robot languages. He also developed CAM-I's geometric modeler, called TIPS (Technical Information Processing System).

The aim of the Japanese is to create an international standard in robotics. CAM-I consists of 48 of Japan's largest firms, including most of the robot builders. In 1983, only 19 United States firms (including the U.S. Air Force) were in CAM-I, as were one representative each from Sweden, the Netherlands, West Germany, and Australia. CAM-I stops short of developing production-type software itself.

Control and software are the two "hottest" areas of robotics at the present. The most radical developments are occurring in robot languages that make simple the programming for complex operations. The two most important current developments in robotics are in off-line programming and speech communication. The greatest problems are still in application engineering and economic credibility. Society demands that industry justify to the worker the reasons for robots.

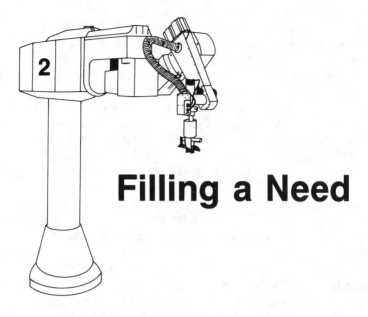

Filling a Need

Industrial robotics is becoming the fastest growing field of manufacturing. *Standard & Poor's Industrial Survey* for 1983 reports that this growth will continue. Electronics, however, is an area almost completely ignored by the robotics industry. (Robots do assemble computer parts such as disk drives, as well as copy floppy disks.) However, the semiconductor fabrication industry is a branch of electronics that robotics has not yet recognized. The semiconductor wafer processing industry, and the printed wire board and assembly activities of the electronic industry require robots with vision.

Today's robot, with its flexible vision systems, was built on pattern recognition work done in the early sixties. Visual robots emerged from the laboratory and were placed on the assembly line only when the solid-state camera was invented in the late 1960s. The solid-state camera utilized the same kind of integrated circuit chips used in the computer industry. Those tiny chips allowed computers, cameras, and robots to be changed as technology evolved. This idea of updating robots with changeable parts allows any machine that is computer-controlled to work faster and become smaller and cheaper. Today every invention in the robotics and computer fields aims at universality.

Electronic inspection is done by skilled workers who look at a semiconductor component and judge the quality of a hybrid, check the pattern correlation, inspect chip selection, judge alignment, detect defects, place parts correctly, and sort wafers. To take over inspection jobs, robots have to be able to interpret "quality." The machines must be able to gauge, identify, recognize, count, and move. Suddenly robots can do these tasks, because they can be equipped with upgraded vision systems. Today robots can hold jobs as inspectors in an electronics facility. Most of the robots involved in electronics inspection work are standard semiconductor capital equipment units which can be upgraded to higher intelligence levels through the addition of vision. Added vision gives a robot speed, complex analytic ability, increased memory to handle large volumes of data, and the ability to extract many features simultaneously. Industry is expanding the capability, independence, and flexibility of robots.

Most market forecasts report that there will be fewer than 100,000 robots in this nation by 1990. A recent Carnegie-Mellon University study predicts that 10 million American assembly workers can be replaced each decade by robots. For robots to move into inspection or gauging, the machines would have to be a cost-effective tool. Right now they are not. For robots to be cost-effective, inspection would have to take place as an added, zero-cycle time-impact—along with the robot's primary location assignment.

The machine vision industry is still fragmented and serving diverse markets and applications. The robot industry depends on the machine vision industry to give robots newer and more difficult tasks able to be performed with simple operations requiring less sophisticated operators. Like computers, user-friendly robots are what industry demands.

Inspection is a task that requires highly developed vision and analytical inferences. Market studies predict ten times more unit sales for inspection vision robots than for the "dumber" servo-controlled robots[1], according to the Robot Institute of America. The whole robotics industry today is copying computers in the sense that the demand for built-in quality is higher than the demand for upgrading dumb, bulky dinosaurs.

Robotics is speeding toward a high demand for vision for in-line inspection of machine tools. The old way was to have matrix and line cameras do inspection or gauging. Matrix cameras are capable of extracting only two-dimensional features. Line cameras focus on sub-mil (i.e., better than 0.001 -inch) accuracy. Vision systems are doing for the robot industry what the microcomputer did for the computer industry.

Vision modules are aiming toward a future where the seeing part of the robot will follow a computer-directed path. This will depend in part on the system's becoming independent of the quality of ambient lighting.

HIDDEN REASONS FOR ROBOT EMPLOYMENT

The old servo-controlled robots must remain expensive because of size, mechanics, and sales volume. Servo robots and seeing robots are branching off in two different directions at present. The way each type of robot is being marketed, the different technology involved, and the system paths are diverging.

Vision is new to robotics. The state of robotics hardware is still primitive. However, the software industry is far ahead of robotics hardware. It is as if the software industry is metabolizing at twice the rate of the hardware side, and as a result the hardware industry is having an anxiety attack.

A robot needs only one type of information to perform its job. A computer has to tell a robot where a part is. The robot asks for location and then goes to rest. If a huge robot is not being used to move from one point to another, the large weight of the robot

---

[1]*Servo-controlled robot* refers to a robot controlled by a servomechanism. This is an automatic control mechanism consisting of a motor driven by a signal that is a function of the difference between commanded position and/or rate versus measured actual position and/or rate.

and the high cost are not space-effective. Back in the 1950s, computers were also not space-effective. Micro-robots are a thing of the future.

Once a robot is used to locate a component and take hold of it, lifting and transporting it to another place, vision then can be added on or built in for such jobs as inspection, identification or gauging, before the robot places the part in its last resting place.

A robot either works in an environment hazardous to humans or performs tasks that require fewer challenges on the job than people will tolerate. However, robots are usually not employed in jobs because people don't want those jobs; instead they are employed because they are supposed to increase productivity.

Robots are on the job because they are assumed to increase quality. They ensure that any defect in craft is uniformly present. Location is everything when it comes to robotics. Robot vision modules come equipped with software and hardware that is programmed to locate the center of certain parts, usually a hole into which a nail, bolt, or pin is driven.

Companies that produce application-driven robots are becoming old-fashioned. Half the robots in the near future will be sighted. In their place are the new generation of seeing robots. As other sensors—touch, speech, etc.—are added, robots can take on more responsibility under the management of a computer which is programmed or operated by a skilled person.

The biggest fear of robotics customers is that there might be too many robots, like too many computers, sitting unused on the production floor because they can't do enough or they constitute overkill. The computer industry estimates that 80 percent of the people who own computers use them for less than 20 minutes a day. In the robotics industry, machines tend to be underused, if the operator is unaware of all the machines' applications.

Recessions and competition have left robot manufacturers beset by customers who want robots that will locate, identify, sort, inspect, enhance profits, increase demand, and make a part better than a person can. In short, if the robot makes smart decisions based on electronic interpretation of visual data, it is in terrific demand.

Today's robots will give answers to the questions what, which, where, when, how good, how many, how big, what+where, and what+how many. By answering these questions posed to the robot by a computer, the robot will proceed to look at a part and give identification, recognition, location, motion, inspection, counting, gauging, sorting, and inventory. The predominant theme in the robotics industry has always been change.

ROBOT RESEARCH IS BIG BUSINESS

After 1980 the robotics industry exploded worldwide as a parallel to the computer industry, while remaining dependent upon it. Robotics is also creating thousands of valuable documents, reports, surveys, manuals, and other literature that is being published at an unprecedented rate. By 1983, the need for a database for obtaining documents on robotics was so urgent that RBOT, a national database on robotics, was developed to capture and catalogue the proliferation of robotics literature worldwide.

RBOT is produced by Cincinnati Milacron Industries, Inc. The database provides information from current literature in all aspects of robotics, from sensor systems to machine intelligence. The database covers both the industrial and business aspects of the industry. Journals, newsletters, conference papers, books, reports, and other sources have been scanned for references to business and technical innovations. The database is available through BRS Information Services via several toll-free numbers: (800) 833-4707, (800) 553-5566 from New York, or (518) 783-7251 from Canada collect.

Other communications in robotics have included videotape films used for instruction available from the Society of Manufacturing Engineers in Dearborn, Michigan, and newsletters on the state of robotics in various countries. *Robotech Japan* is a monthly newsletter on Japanese robotics developments and is published by Yoshiko Enterprises in Winter Park, Florida. Numerous robotics publications are opening each month to meet the increasing need for information.

In Washington, DC, The Robotics-Automation Network (TRAN) and the Robotics Roundtable are new organizations that exist solely to gather and disseminate information on human factors in automation. The organizations scan conferences and workshops for information, but their key link is the Washington-based American Association of Community and Junior Colleges. The roundtable approach is focused on the study of training, technology transfer, economic development, jobs, and foreign trade in the robotics field.

Everette/Charles, Inc. (Fontana, California) has set up a new business for telemarketing, direct-mail, and catalogue promotion of factory automation hardware. There has been a tremendous increase in businesses that do research for the robotics industry and sell information.

ANALYSIS OF ROBOTS ON THE MARKET

Technology is universal. Laser-equipped robots and other laser-utilizing machine tools are used to perform various metalworking tasks in unison instead of sequentially. In October, 1983 the Japanese government unveiled its small, prototype factory of the future—a $60 million project that cut production time in half. By 1988 this laser-run-in-unison factory will be available commercially. The production tools are being developed at the present time.

The lasers performed on-line heat treating, surface alloying, welding, and deburring. American arc welding equipment manufacturers currently are teaming up with Japanese robot arm suppliers to produce the latest arc welding robots. There are only 300 arc welding robots in the United States, but 3000 can be found in Japan. By using robots as welders, U.S. manufacturers have projected that an arc welding system can be operational 70 percent of the time instead of the 30 percent that it is at present. Behind this research are the robotics laboratories at Massachusetts Institute of Technology, Carnegie-Mellon University, the University of California at Berkeley, Ohio State University, and the University of Wisconsin, according to *Iron Age,* a metalworking industries trade journal.

ROBOTS IN PUBLISHING

Robots are taking over the printing, publishing, and graphic design industries in the United States, according to a June 1983 article in *Robot/X News.* These industries consist of 50,000 businesses with an estimated 1.3 million employees, according to recent industry figures. Publishing has been suitable to robotic automation because it is both labor- and skill-intensive. The publishing and printing industries are based on the economic reality of immediate return on investment.

Web printing presses have reached high degrees of automation without robots in such computer-controlled operations as newsprint transportation and splicing rolls of paper without stopping the press. Remote-controlled operations such as controlling the mixture of water to ink has been computerized for a decade. Suddenly robots have been introduced into large-scale newspaper production. They have automated the mail room operations. Robots insert, sort, and convey tied bundles to trucks. The first robot to enter publishing came in by way of newspaper publishing.

A robot from W.M.I Robot Systems, Inc., Hialeah Gardens, Florida, was employed in the delivery end of a multi-web press at the George F. Valassis Company in Livonia, Michigan, which turned out over a billion free-standing newspaper inserts for over $100 million in sales in 1982. W.M.I Robot Systems has robots working the binding operations at R.R. Donnelley Company and at the Salem, Illinois, plant of World Color Press. The robot handles finished packages of books, palletizing the stacks at the rate of 320 books per minute.

In a variety of printing businesses robots are called *labor intensive*. The robots prepare materials for the press. Computerized typesetting is still performed by a person operating a keyboard in the manner of a typist. Automation is taking a different direction by creating voice character recognition systems by computers for typesetting input.

The roboticized press is still in the near future, but is coming quickly. High-circulation magazines compose complete pages of type, photographs, retouching, and corrections using electronic pagination systems that have replaced hundreds of retouchers, paste-up artists, and page assembly technicians (strippers). Instead of all these employees, there now stands one electronic station. One operator sits at the console and another stands by the electronic scanner that digitalizes the photo images.

In Japan the press is already operated mainly by robots. Tokyo Kikai Seisakusho, Ltd. (TKS) and about 150 other publishing firms in Japan use robots in production. In the United States, however, newspapers and books run in much less standardized fashion; printing is much more in a spur-of-the-moment operation. A customer enters with a rush job one day; the next day could be slow.

Publishing requires the highest flexibility, and robots were designed to be adaptable. A web press designed for publication work can handle all sizes of published material. Robots are creating the need for standardization in the printing industry. The variety of the sizes of books would have to change. Robots are entering wherever manual labor is done, particularly in the post-press operation. Robots carry tons of papers coming off the presses. Robots do binding, trim book blocks, and apply book covers.

Material-handling robots are required to adjust to a volume of work that changes daily. Where part-time, unskilled help once was hired, now robots are keeping stacks of materials moving.

A robot in the publishing industry palletizes magazines, bound books, directories, and catalogues. It handles logs of press signatures, which are untrimmed sheets containing a dozen or more different book pages. Such a robot costs at least $70,000. In addition, the robot needs peripheral equipment such as conveyors and automatic palletizing units. Such a robot frees at least two employees for other operations and is cost-effective in high volume printing operations.

In addition to the W.M.I. Robot, two other publishing systems robots are used in graphics or printing. One is produced by Franchise Mailing Systems, Milwaukee, Wisconsin, and another robot (for graphics) is made by Didde Graphic Systems, a forms press manufacturer. Unrelated industries such as printing, fashion design, farming, and homebuilding are a far cry from automotive and electrical machinery, yet all of these industries are buying up the robots at varying speeds.

WHO IS USING ROBOTS?

Robotics companies are multiplying rapidly. To analyze today's robots on the market, a survey was conducted among Japanese companies who will use robots in the future, as well as those who presently utilize them. The reason behind this logic is that whatever Japan does today with robots, the U.S. is predicted to be doing in 5 years.

Participating companies received a questionnaire analyzing their current stock and estimated inventories for 1985 and 1990. The results revealed that the automative and electrical machinery firms were the heaviest robot users. Other active firms included the metal processing industries and the plastic fabrication businesses.

The Japanese fashion design/apparel industry is in heavy competition with those elsewhere in southeast Asia because labor and parts are cheaper outside of Japan. Recently a 7-year project to develop fashion designing robots that will also cut, sew, finish, and inspect clothing began in Japan. The Ministry of International Trade and Industry will spend $65 million for robots with sight and touch integrated with computerized design and cutting systems. In 1985, Japan is to start using robots in the following non-manufacturing jobs: insecticide spraying, egg inspection, fertilizer spreading, and cargo handling. By 1990, non-manufacturing robots in Japan are projected to become a $360 million industry that will employ robot nurses, guides for the blind, and street sweepers, according to the *Industrial Automation Reporter.*

An advantage to having a robot with planning ability (artificial intelligence) is that instructions to the robot can be simple. This point is the key to the explosion of the robotics industry. The results of the robotics demand forecast sent to Japan revealed a percentage ratio based on the type of application used by robots. The survey did not apply only to Japan; it revealed the trend of the worldwide use of robots. The forecast showed that the use of robots by Japan equaled the estimated nationwide average.

According to the Japan Industrial Robot Association, the robot survey of that nation showed that over a 10-year period, the number of robots will grow substantially. At the end of 1984, 570 organizations doubled the amount of robots used in 1979. By the end of 1989, the robots in use will double again. That is, there were 26,942 robots in use in 1984; there will be 43,595 robots in use by 610 organizations in 1989. Just a few years ago there was a worldwide population of only 30,000 robots.

The growth of technically advanced robots such as variable sequence and playback robots is expected to rise. The number of playback robots is predicted to increase 10 times within a decade. The growth of fixed-sequence robots may be slow.

The statistics are said to be underestimated because of a lack of response to the survey. Also, the number of new robots acquired isn't represented in the survey. Currently the cost of a robot is higher than that of labor and will continue to be that way until 1985. By 1990, robots will become cost-effective because labor cost per person will exceed robot cost by approximately $10,000 per person per year for Japan alone. There the investment rate also surpasses labor cost. The turning point in Japan is expected to be 1985, when robot demand is estimated to be 43,000 to 55,000 units, with increases up to 94,000 units by 1990. The survey estimates were very conservative. How this will affect robot demand in the United States is still uncertain.

Robot applications in the United States promise increased production—but only up to 30 percent. One servo-controlled robot costs $40,000. Two non-servo robots cost $36,000 ($18,000 each), and the buyer avoids the cost of safety equipment—as much as $20,000 per machine. Scrap is reduced by 20 percent. A robot can pay for itself in 9 months to 2 1/2 years. In the United States, a survey of welding engineers by *Welding Design & Fabrication* indicated that under 10 percent used robots. Over 25 percent surveyed said that they planned to use robots for welding by 1989. In the United States robots are being used to manufacture computers and software and even to copy software disks.

Robotics industries tend to cluster around intellectual centers. Labor must be cheap, stable, and skilled, and the robots must be versatile. Jobs in robotics have to attract high-level managers, so location must be suitable for people. Capital for robotics development must be available from financial services. All these factors will act in unison to control the expansion of robots in this country.

ROBOTS PHOTOGRAPH TELEVISION COMMERCIALS

According to "Robots Make T.V. Commercials," a June, 1983, article by Dan Abramson in *Robot/X News*, people who create the special effects in television commercials are using robots as part of the production techniques. Robotic boom arms attached to cameras photograph moving objects while the camera is in motion. The robot has accuracy to 0.001 inches. No human arm can hold a camera with such steadiness.

Camera operators are being trained to use robots. In fact, many of your favorite television commercials have been filmed by robots that get it right on the first take. Opening credits, logos, images of stars swirling in the blackness of space, a rotating galaxy, all are images created by robotics technology. Each star moves past the camera several times while the letters that spell out the opening titles move into place on-screen. A platform-size animation stand is hooked to a computer.

The entire animation scene is programmed into the computer, which can replay the scene as many times as required. The use of robotics allows the camera to go back to certain positions. Effects are created by robotics that could never be done manually. The robot repeats the camera moves at a certain speed, working consistently, never deviating by a thousandth of an inch. Everything moves in tandem with the robot arm holding the camera. The robot allows a television commercial to be made using a system that is flexible and repeatable. It is computer-controlled stop-motion photography.

Multiple runs of the same images can be done with a robot and computer. Software is being constantly updated that makes it cheaper to use a robot to film television commercials. Before robots were used to film commercials, a studio had to hire computer programmers. The technology to move objects by computer has been around since 1974, but the software wasn't developed yet. A decade later the software is here, but the robot replaces the computer programmer.

The first few seconds of many commercials consist of three-dimensional animated logos that can be created by robots photographing light sources in space. Robotic multi-images are used to coordinate continuous motion shots with stop-motion photography.

A robotic camera system allows a television commercial producer to complement other capabilities, such as computer-assisted animation, live action, computer-assisted optical printing, and computer-generated imagery.

The robot's boom arm integrates with other film techniques and works in three dimensions. Robotic technology makes it possible to shoot the same model against different scenes while allowing different shots to match up identically.

Robots broaden diversification in almost any field they enter. They allow a company to do jobs in more imaginative ways and still meet the budget and deadline.

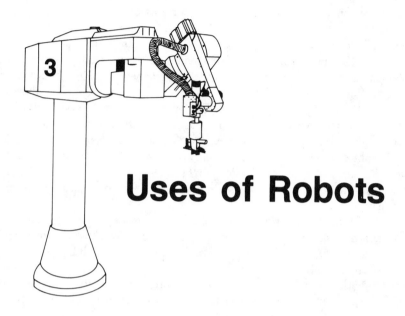

Uses of Robots

Robots fuse the knowledge of many experts into one machine tool. They have been created to perform tasks normally done by human labor, working as surrogates for human workers. They don't require great technological sophistication of the user. Robots work tirelessly in any atmosphere. In their 40,000 hours of working life, robots can lift heavy loads, reach long distances, and perform repetitive tasks without errors due to distraction. These machine tools are used for arc and spot welding, paint spraying, glass handling, heat treatment, forging, machine loading, stamping, plastic injection molding, investment casting, conveyor transfer, palletizing, inspection, home security, and serving the needs of the handicapped. Promotional and "home" robots can also act as personal servants.

In the military forces, robots are being tested to handle dangerous jobs. Robots actually fight on the battlefield, guard and defend nuclear storage facilities, and perform sentry duty. These machines transport and sort heavy ammunition and disarm live bombs and shells.

Robots fight fire on aircraft carriers. The Navy's Robotics Research and Development Laboratory in White Oak, Maryland, is building a prototype robot for aircraft carrier fire control. The Army is developing robots that defuse bombs and patrol battlefields in the face of toxic chemical weapon attack. Robots are immune to biological warfare, nerve gas, and other various types of weapons used against humans. In a more practical sense, the Navy uses robots to scrape and repaint ships. Humans who used to be assigned these jobs are now being trained for technical jobs that require more skill and less muscle.

The Defense Department and the Navy have a robot that finds and indentifies rivets.

It pulls out the rivets from curved surfaces on aircraft, drills and inspects the holes, marks flaws, and logs every related piece of data. The robot eventually will disassemble the aircraft. This "de-riveter" robot has been identified by an MIT panel in 1983 as the number-one robotics project in the world in terms of technical challenge and potential.

The rivets in aircraft skin are usually polished smooth for aerodynamic reasons. A person can see them, but for a robot to find them and then accurately de-rivet requires a new sophistication in robotics technology. Disassembly of corroded aircraft wings has cost the U.S. Navy a large sum of taxpayers' money, money which this robot can save. The Naval Supply Center in San Diego, California, was awarded $2.4 million in 1983 in a two-year contract with Southwest Research Institute in San Antonio, Texas, a not-for-profit contract research, development, and engineering organization noted for engine fuel and lubricant development. The contract allowed the Naval Supply Center and Southwest Research Institute to develop a mobile robot that drills out rivets at the rate of 120 per hour. Robots have to remove thousands of rivets to get at internal components that may be corroded.

This de-riveter robot has a critical impact on the worldwide robotics industry. Its vision system and method of operation allows similar systems to line up with semi-randomly oriented parts. A robot like the de-riveter could speed the robotics industry light years ahead in assembly tasks.

The entire robotics industry depends upon developments in control and sensory software. Without programs to run the robot, the hardware is just a chunk of metal. Computers manage robots and must interact with them.

A computer programs a robot to handle differing circumstances on one job. If the programmer's hand shakes, the robot will shake until it is reprogrammed. A robot will imitate the movements of the programmer with many times the strength and endurance.

ROBOT OPERATIONS

The robot locates a component, such as a rivet. The robot is "put to rest" while a mounting device with tactile sensors probes the surface of an object until one robot finger touches what the robot is programmed to seek out. The robotic platform then orients itself perpendicular to the object's surface so a task can be performed. For example, a drill bit can go into the rivet.

Computers that direct these robots are conventional minicomputers. Personal or "hobby" robots are run by microcomputers. (One significant exception to this generalization is shown in Fig. 3-1.) Computers control the activities of the robot's microprocessors associated with the tools and sensors; robot operations require a high degree of accuracy and repeatability, which is provided by the controlling computer. (The position and operation of the IBM robotic arm shown in Fig. 3-2 is monitored 50 times per second.) Human workers are programming the computers that manage the robots. A robot can perform the same job underwater, in space, deep inside a mine, or in any other environment that would be hazardous to people.

Robots are systems that can act as intelligent assistants. These machines have sorting ability. A robot programmed by artificial intelligence and controlled by computers will pick out specified components from randomly strewn parts. This is called *silhouette vision*. If the robot is equipped with this type of sensing, it will put the components

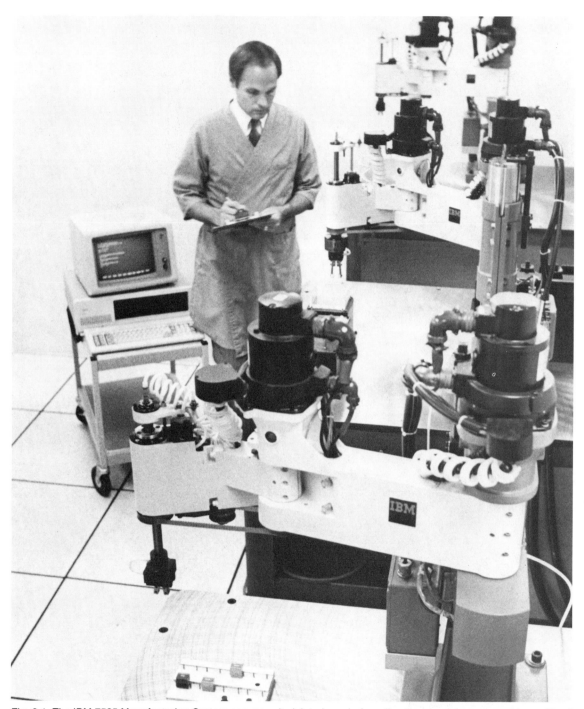

Fig. 3-1. The IBM 7535 Manufacturing System can move its jointed arm in four directions to pick up, assemble, and load with great precision parts such as those shown in the foreground. This robotic system can be programmed with the IBM Personal Computer, left, using a special version of the IBM-developed robotics language, AML. Courtesy International Business Machines Corporation.

Fig. 3-2. The fluid, accurately repeatable motion of the IBM 7565 Manufacturing System's robotic arm is captured in a multiple-exposure flash photograph. An IBM Series/1 computer monitors arm parameters 50 times per second, allowing the system to identify and correct problems rapidly. Courtesy International Business Machines Corporation.

in a specified order. The robot needs the darkest background to "see" contrast; the parts have to be silhouetted. The computer program allows the robot to measure the brightness between the glint of light bouncing off the components and the blackness of the background. This value enables the computer to see the difference between the parts and the background in order to perform *outlining*.

The robot's system uses contrast as a guideline. The bright components glow on a black field and the robot outlines the parts. The system goes to the borders of the silhouette. Television scanners help the robot draw the edges of all the parts. The robot has to select the correct component; to do this a computer acts as the robot's brains and finds the "geometric center of gravity" of each part's silhouette. This is usually the center of the silhouette, as well as the midpoint of the part's ellipse. The computer controlling the robot finds the sizes of the parts as one part relates to another. For each part the computer creates an ellipse equal in area to a particular component.

The orientation of each part is determined by studying the tilt of the component's ellipse with respect to the computer program. From this point the robot continues to be programmed by a computer, which refers to a checklist program written in logical sequence for all parts to be identified and sorted. Every part is given a number of points called *weighting factors*. The points for each part must add up to a certain value before a part can be identified and sorted.

Robots can act as consultants with inference and learning mechanisms of their own. These machines can be connected to worldwide databases for the expertise necessary for making important decisions. Robots enable people to go on to higher challenges in the work place. Outside of the manufacturing industry, robots are used for maintenance duties. A floor-cleaning robot now in use in supermarkets has been tested by the Navy to scrub the decks of ships. An Air Force base in Oklahoma City uses robots to work in aircraft engine maintenance. Robots are being "drafted" by the military forces because between the present time and the end of this decade, personpower in the armed forces is predicted to shrink by 25 percent. Robots will do the most mundane chores without developing morale problems; a robot's retention is unvarying.

When people are introduced to the concept of robotics, they tend to flash back to images of past horror films—the giant, evil robot policeman in *The Day the Earth Stood Still* or Dr. Frankenstein's ubiquitous monster. When robots are built to resemble humans, people either fear them as "the unknown", or tend to talk to them as if they were alive. Industrial robots tend to look like machine tools—which they are—while personal or promotional robots are usually built to look child-like or vulnerable. Because robots lack emotions, they are assumed to have the capability to be cruel or somehow take away something of value from people. But it is the human programming the robot who gives the machine its qualities.

In the film *Star Wars*, the robot R2D2 is no threat because it is as powerless as a pet. The fear of giving robots power over humans has colored the historical image of robots in the media. Humans recognize the limitations of their own physical and mental boundaries, and there often is a not-quite-rational feeling among people that robots are superior because machines can reach into the most inaccessible places and do better work.

ROBOTS IN TRANSITION

Fritz Lang's world-renowned 1927 film, *Metropolis*, depicts a scientist who creates a robot that looks like a young woman. The film is about a future war between labor and management. Industry, controlled by a strong boss, is frustrated with human workers' productivity rate. The boss wants to fire the oppressed workers and create robots that will never become exhausted. The film is about assembly-line morale.

In *Metropolis* the industrial engineer is depicted as an "insane" man seeking money from a rich industrialist. The designer who is asked to build a mechanical robot instead builds a *fembot* (female robot) and gives it secondary sex characteristics. The fembot becomes a liberationist who leads blue-collar employees in a revolution. The fembot eventually is smashed to worthless chunks, along with the rest of the city, while the strong boss retires to his suite to seek other ways to improve his profits. The film script was written by Lang's wife, Thea von Harbou. Lang and Harbou created a familiar media image of factory automation on the assembly line. The message was (and is) that we fear our own limitations and project the fear of loss onto a machine. The assembly line represents the conflict between boredom and security that maintains a hold on the laborer. The robot on the assembly line becomes the unknown fear.

Once the robot in film actually made the transition to the robot in industrial reality, discussion of robotics turned to profit-taking. Approximately 85 percent of all manual labor expended in industry in the nation is assembly work. In the United States, studies have shown that $900 million per year was being spent on assembly line tasks. Of this,

$600 million was spent for jobs that could be handled by robots. Of that $600 million, half was being paid out for jobs which could be handled by obsolete robots that were standard machines back in 1980.

The capabilities of computers to provide information has been increasing the desire for information at an exponential rate. Industry is riding a trend to replace capital equipment for human workers. Only 30,000 robots populated the world in 1979. In 1984, 570 companies in Japan alone purchased 26,942 units, doubling the number of robots (13,120 units) that Japan purchased in 1979.

Some 20 years after robots have been introduced into the workplace, the robotics industry is still in its infancy. At the start of this decade only 4,500 robots were sold in the United States, producing a total sales revenue of $100 million. By the end of 1982 that figure leaped to $200 million—doubling in a year. Recent forecasts project annual sales revenues for robots to reach as high as $3.2 to $5 billion by 1992. This figure may be compared to the present $157 billion worldwide computer industry, which has preceded robotics by only two decades, to predict continued growth in robotics.

Due to the financial success of the computer industry, abilities of robots are being extended by the advances in computer *artificial intelligence,* a term meaning planning performed by a machine. Artificial intelligence is the mode of programming that allows a computer (or robot) to operate on its own, to learn, adapt, reason, or correct and improve itself. The trend is for robots to become faster and more intelligent in a technological "conspiracy" that is unparalleled in human history.

A manufacturer of remote-controlled entertainment robots, 21st Century Robotics, plans to jump into building its own robot for police work and firefighting, according to a March 23, 1984, report in the industry newsletter *Robotics Insider.* The machine is called the Tactical Support Robot. This company shows a trend toward the manufacture of military, police, and firefighting robots as it joins other firms such as Denning Mobile Robotics (prison security) and Robot Defense Systems (battlefield sentries) to create the security robot.

In the robotics center at Georgia Tech, 21st Century Robotics is developing a research and development team. A direct-mail campaign reached 9100 U.S. law enforcement agencies to make them aware of the TSR robot that has a base price of $20,000 to $45,000. The TSR robot has four functions: handling bombs and unexploded military ordnance, hazardous materials, SWAT-team type activities, and remote firefighting. The firm's source of capital is internal, from sales and rentals of promotional/educational robots. Rentals (including the operator) bring in up to $750 a day, a highly profitable venture. Coca-Cola is one of the users of these promotional robots.

Industry uses robots as machine tools. However, robots have graduated from repetitive, dirty, work and are now ready for new challenges. Economics, i.e., trade, always inspires new ideas. This in turn foments revolution. People engaged in trade simply want to make profits. Today robotics is a potent tool used to accelerate the diffusion of knowledge. The field of robotics is charged with drama and emotion, especially to blue-collar workers who are turning from unemployment to retraining in robotics and computer technology.

Like the invention of the computer, computer application to robotics has changed the way people think, work, and live. We stand before a singularity, the robot, a machine so unprecedented that predictions are still science fiction. Reasoning machines are changing life in unpredictable ways. Who is to say how access to robots will challenge the development of the quality of life? Robots have universal access to machine intelligence which will one day outstrip human intelligence by its speed and depth.

As long as there is money for the development of software technology, robotics (and hardware) will progress at a rate that may test our adaptation to rapid change. In the following section let's take a look at the robots on the market to see how they are put to work to regear our lives.

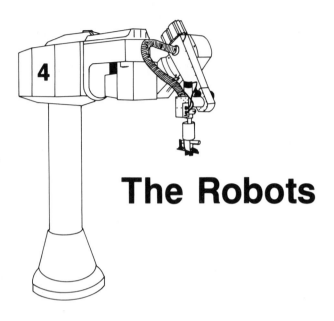

The Robots

In this section you will find out who the manufacturers are, the specifics of the various robots, location, industries, users, and a summary of the robots' primary uses and significant selling points. Robots can be used for entertaining, promoting, securing, battlefield fighting, firefighting, assembling computer components, making cotton candy, welding, painting, sorting, de-riveting, and loading—to mention just a few jobs.

The type of robots represented in this chapter are diversified. You will find that robots come in many different sizes, shapes, prices, and uses. The *Star Wars* overtones just won't go away. The difference between a robot and an automated storage/retrieval system is great. This distinction is often lost on the public. A robot can be taught to adapt itself to change.

The reaches of the human mind are realized in hard metal and baseline profits. Carnegie-Mellon University's Robotics Institute is building and testing a stripped-down mechanical dog to extend its work in understanding the act of walking. The 3 feet by 2 feet robot puppy is under construction in the university's "Leg Laboratory" in Pittsburgh, Pennsylvania. It is efforts such as these which find a practical use for the imagination. In this chapter let us take a look at some of the robots that have changed the world.

ARMSTAR

The Armstar robot is manufactured by Tokico America, Inc., of Dearborn, Michigan, and Los Angeles, California. This $150,000 painting robot, introduced in 1978, is used worldwide for automating the toughest finishing tasks in such industries as automotive, electronics, appliance, and plastics manufacturing. The robot is geared for the total automation of assembly-line undercoating, sealing of weld seams, interior and exterior prime or top coating, and adhesive application.

There are over 400 Armstar robots in operation around the world. Unique features of the Armstar are high-speed optical scanning and a microprocessor-controlled "pathminder" finishing system that operates faster than any other continuous-path robot. The robot also has a self-diagnostic system and features easy program editing.

Robot Description

The Armstar robot system components consist of a manipulator, hydraulic power supply, microprocessor control console, and modifier console (Fig. 4-1). The work-supply device contains two carriers with a rotation control panel for automatic operation and two operator control stands. The Tokico painting robot will repeat painting operations continuously and automatically according to memorized programs. These programs are taught to the robot by either the point-to-point or continuous-path methods.

In the *continuous-path method*, the worker teaches the painting operation by moving the robot arm as he or she usually handles a spray gun. The *point-to-point* technique allows the worker to separate the movements of the robot arm into linear motions, teaching only the starting and ending points of each motion.

The robot has a flexible wrist that can paint narrow, inside, turn-up, and hidden parts of complicated work pieces such as automobile bodies. The wrist of the 6-axis robot moves vertically, horizontally, and around like the human wrist, allowing, the painting of the inside of a box. Low-noise electro-hydraulic power operates the manipulator. Conventional floppy disks are used to store the painting programs; data forwarding between the auxiliary memory and main memory is carried out with the push of a button. The robot stops at operation errors or malfunctions, a buzzer sounds, and a numeric error code is shown on the control console to indicate the cause of the trouble.

The 5-axis robot has 70 degrees freedom of movement of the arm, 210 degrees of movement in the wrist, and 90 degrees pattern rotation. In the 6-axis robot, the arm movement is 100 degrees horizontal, 75 degrees back and forth, and 70 degrees vertically. The freedom of movement of the wrist is also 210 degrees, and the wrist rotation is 260 degrees. The 6-axis robot features 210 degree pattern rotation. The robot has six elements that can be modified: position and angle at each type of movement, spray on and off, speed of the robot, increasing or decreasing teaching points, back-and-forth or continuous sweeps, and selection of spray patterns.

Industries and Applications

The Armstar, also known as the Tokico painting robot, is targeted for use in spray painting or otherwise applying coatings to objects of all sizes. Six of them can be grouped in three paris to form a tunnel robot, each pair being mounted on a common slide so that they can move horizontally. The arms are then oriented to form a tunnel, with each arm able to extend vertically and articulate at the wrist. In the paint shop this team of robots can apply weld sealer or polyvinyl chloride, underseal, primer, surfacer, or color coats to large objects such as assembled automobile bodies. It is with the primer/surfacer function that the Armstar has made the biggest advance; it can be used to apply adhesive to automobile windshields—even ones with compound curvature—prior

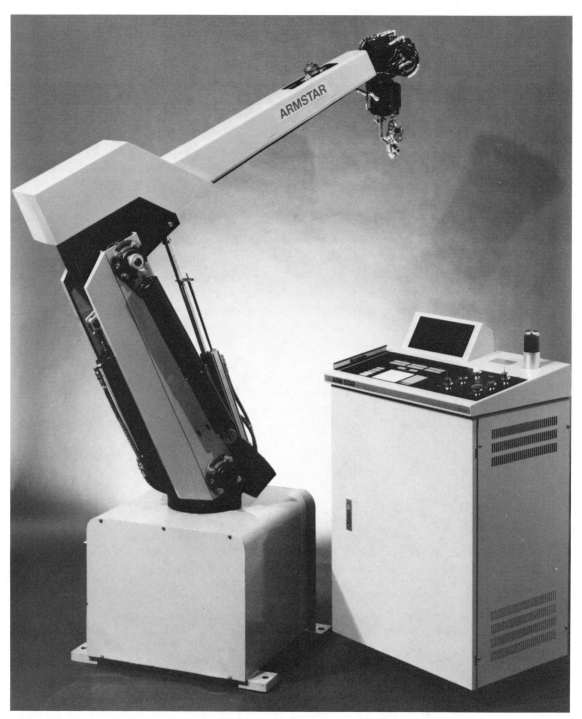

Fig. 4-1. The Armstar spray painting robot, shown here with its controller cabinet, is used worldwide for diverse applications such as automation of assembly-line undercoating, sealing of weld seams, interior and exterior prime or top coating, and adhesive application. Courtesy Tokico America, Inc.

to direct glazing. This robot also can be set up to weld at junctions.

The Armstar robot can accomplish the work of two or more painters on one shift; alternate work stations are within easy reach of an operator. The compact design of the robot allows installation next to existing plant conveyor systems, and it be placed overhead or on floor-mounted delivery systems. The user can select a large- or small-size robot, depending on the painting booth space and the size of the work pieces. Tokico will lease the robot to minimize capital outlay; though service contracts are not available, the unit also comes with a 1-year warranty.

In Hofu, Japan, a new Toyo Kogyo assembly plant produces 20,000 Mazda 626 sedans per month with 1800 employees and several Tokico robots. The primary innovation in this industry is the tunnel robot mentioned earlier, developed jointly by Tokico and Kawasaki Heavy Industries. The tunnel robot is set up for welding, making 33 or 35 welds between the front, center, and rear floor panels of the Mazda chassis. It is a compact installation; only one control unit is required for 62 functions and 62 axes of motion.

In the paint shop, Toyo Kogyo utilizes 20 Tokico painting robots in various applications. For example, in the primer/surfacer booth four robots working in conjunction with two Minibell electrostatic spray systems apply all the paint. A pair of simple robot arms open and close the car doors for the painting robots, which can spray the inside of the car body, including the door hinge pillars.

Users and Operations

The sequence of operation begins with the operator placing the part to be painted into a carrier and pressing the "in" button; the carrier then slides into its root position. Upon receipt of a signal from an in-position carrier, the robot starts the spray cycle. When one side is completely sprayed, the robot signals the carrier for a 90-degree rotation, and then continues the spray program. The process continues automatically through each rotated position. Upon completion, the carrier returns to the operator load area for removal of the finished part, while the robot moves to the adjacent carrier for another spray cycle.

Any part of the robot can be programmed via a lead-through teaching method, either in continuous-path mode (for easy tasks) or in point-to-point mode for tough, high-speed jobs. A balance mechanism and reduced weight of the robot's moving parts makes teaching relatively effortless for the human instructor. Any mistake made in the teaching process can be corrected readily with the modification console. The start timing of teaching and that of playback are synchronized automatically by an electronic circuit without further adjustment.

In the Toyo Kogyo plant, a worker can be trained quickly to operate the robot; once programming is complete, operating the robot is easy. The automatic program selection device option allows the operator to enter the program number for the particular task(s) into a key-in input device for transmission to the control console. If different types of objects are supplied in a given order, the "digi-switch" method allows the operator to set the program numbers for the objects in the order they are supplied.

The Toyo Kogyo robots are installed in pairs, opposite one another, and there is a third pair further down the line to spray matte black paint (wet paint on top of wet paint) on some models. Two lines exist in the color coat booths, with two robots and five humans in each booth. According to *Automotive Industries* magazine, the presence of the workers to some extent negates the advantage of the robots, but engineers point out that each robot in the paint shop replaces one human per shift. For a two-shift workday, 20 robots do the work for 40 humans.

Summary

Name:	Armstar (Tokico Painting Robot)
Manufacturer:	Tokico America, Inc. 15001 Commerce Drive North Dearborn, MI 48120 (313) 336-5280
Physical Description:	Horizontal arm attached to vertical arm on rectangular raised base. Paint-spray gun can be attached to end of arm. Also can be programmed to weld. Weight is 1200 pounds.
Primary Use:	Spraying paint. The robot is a finishing system for nonconveyorized shielding and masking operations.
Significant Sales Point:	The robot continuously and automatically repeats painting operations according to memorized programs.

ASEA

The ASEA Group of Vasteras, Sweden, is a firm specializing in a wide variety of industrial goods, ranging from railway equipment (ASEA's sleek AEM-7 electric locomotives, made in the U.S. by General Motors, hustle Amtrak's *Metroliners* between Washington and New York at over 100 miles per hour) to a well-known line of arc welding, deburring, gluing, materials handling, and assembly robots the company has manufactured in the United States, Sweden, and other European nations for over a decade.

ASEA (not an acronym, by the way) built its first all-electric, microcomputer-controlled robots as long ago as 1973; these machines have been in nearly continuous service since then, working through three shifts with an average of 98 percent "uptime." The company currently is working to broaden the sensing abilities of its robots through tactile recognition and simulated vision systems, and has exhibited vision-equipped robots which promise to reduce programming time by at least 25 percent.

In the computer industry, computing costs have decreased by 20 percent per year since 1954; robotics is providing similar cost-performance ratios. Although products made by robots are less costly than their human-manufactured counterparts, many U.S. industries are finding that declining revenues have nonetheless limited the capital resources available for purchasing automated machinery. Despite this somewhat mixed economic picture, Paul Brunner, president of Milwaukee-based ASEA Inc., expects continued fast growth for ASEA robotics.

Robot Description

ASEA robots all bear a striking family resemblance to one another, looking a bit like squarish orange fire hydrants with two extended arms—one vertical and one horizontal (Fig. 4-2). Let's examine a few of them now.

The IRb 6/2. The quiet, electric-drive ASEA IRb 6/2 robot (Fig. 4-3) is designed as an all-around process system capable of operating in difficult environments. It consists of a control cabinet with portable programming unit, a measuring and servo system, and the robot itself. The lightweight, servo-powered arm accelerates and decelerates rapidly, can lift a 13-pound load, and has a repetition accuracy better than ± 0.008 inch. The robot's six degrees of freedom include rotary movement, radial and vertical arm movement, rotary and bending wrist movements, and optional horizontal travel.

Teaching the IRb 6/2, as with all the ASEA robots, involves a combination of tools and methods: an alphanumeric display using plain English, a series of touch-sensitive buttons, and a proportional-speed joystick control lever. Complex programs for multi-axis robots can be compiled with dissimilar items. Different operating sequences for items, addition and omission of steps, integration of the whole operating sequence, and additional variations all can be accommodated easily.

The IRb 60/2. ASEA's heavy-duty IRb 60/2 robot handles larger payloads (up to 132 pounds) and has its components enclosed for protection from hazardous environments. An advanced microcomputer control system, coupled with all-electric drive, allows it to repeat complex sequences of motions with an accuracy of ± 0.016 inch.

The IRb 90S/2. This robot, ASEA's spot welding specialist, can be used to handle the heaviest payloads under the most severe working conditions. Servo-controlled d.c. motors and special software give high accuracy in the short, rapid movements

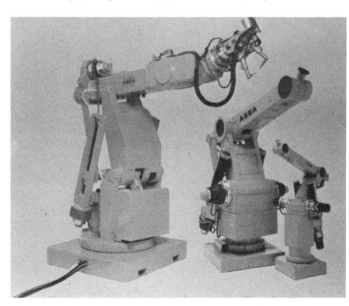

Fig. 4-2. Family portrait of the ASEA IRb robots. Shown here are the 6/2, the 60/2, and the 90/2. Courtesy ASEA Robotics, Inc.

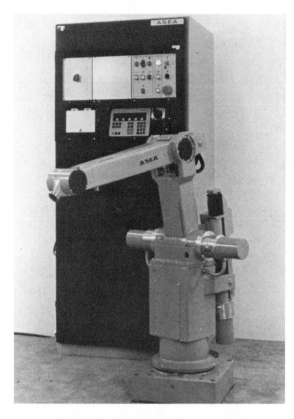

Fig. 4-3. The ASEA model IRb 6/2 robot with controller. Courtesy ASEA Robotics, Inc.

characteristic of automated spot welding; enclosing water, air, and power supplies inside the robot arm allows it unrestricted use of its work envelope without hanging cables. Programs can be altered and edited after they have been stored—and even while they are running.

Implicit in the IRb 90/S2 is the ASEA philosophy that specialized robots should be considered only if the machine will be dedicated to the same general application, with variations, throughout its entire working life. Since virtually all of the software, tools, accessories, and interfaces have been worked out by trial and error, first-time robotics users are advised to start with a "naked" robot as part of a fully developed system.

The MHU Senior. This machine is a pneumatically powered industrial robot that can lift loads up to 33 pounds. It consists of an arm, rotation unit, and column, all provided with quick-release electrical and compressed air connectors for easy installation, dismantling, and transport. The arm has a 43.3-inch stroke and eight stop positions; the rotation unit has six stop positions and can turn through any number of successive 360-degree rotations in either direction. The robot has 19.7 inches of free vertical movement on the column, with four stop positions. The MHU Senior is regulated by the PC 80, a microcomputer-based control system capable of simultaneously controlling several robots or other machines performing different tasks.

The MHU Minior. ASEA also manufactures the MHU Minior. This is a pneumatic pick-and-place robot capable of rapid operations in applications with short transport distances, such as with small presses or automatic assembly machines. Its arm movements are limited vertically and horizontally by damped, adjustable mechanical stops; arm position is monitored by inductive sensors. Gripping is performed by one of ASEA's standard gripping modules, either fingers, magnetic grippers, or suction cups. The MHU Minior normally is teamed with the microcomputer-based PC 40 controller.

An important feature of ASEA robos is *adaptive control*, a relatively new, feedback-oriented control system in which one or more sensors provide information to the robot to modify and refine its movements. Adaptive control provides a robot with a sense of "touch" from which it can "learn" on its own—giving it, in effect, the ability to cope. This technique requires less critical accuracy during initial programming, fewer programmed points for a particular task, no programming for minor position changes, and makes fewer demands on the equipment.

ASEA's adaptive control mechanism confers on its robots both point adaptivity and curve adaptivity. *Point adaptivity* is a type of programming that gives the robot several different ways of searching for points that are not necessarily determined in advance; positions of programmed points can be corrected automatically. On the other hand, *curve adaptivity* allows the robot to work along one or more curves whose shape is defined by information from a sensor. (This capability was added to ASEA robots in 1979.) In both cases, adaptive control refines the programmed pattern and speed of movement in response to changes in dimensions and location of the work pieces, as well as changes in the work place.

Industries and Applications

Although ASEA's robots are targeted for use in a variety of applications, the steel fabrication, welding, and assembly industries have proven to be their biggest users. An ASEA robot can be used to handle massive, heated parts at presses and forging machines. Separate loading and unloading through the short sides of the press make for high productivity, efficient material flow, and easy access to the press for tooling changes.

Practically any part that can be welded manually can be spot- or arc-welded by a robot, within a mildly restrictive set of conditions to which ASEA robots are well adapted. (ASEA targets its welding robots at companies that employ three or more welders and use 20,000 or more pounds of wire per year.) Steel ranging from thin sheet to heavy plate is the most commonly welded material; in most cases, the steel welded by robotic systems is 3/16 inch or less in thickness. Aluminum also is being welded successfully in robotic systems, as are stainless steel and alloys.

The weldment must be small enough to fit within the working range, or *envelope*, of the robot and its manipulator. As a rule, parts to be welded in robotic stations should be no larger than an average office chair; larger parts can be handled with positioning equipment or by mounting the robot upside down over the work area. The robot also can be placed on tracks or rails so that it can move over extended distances to weld long seams and reach extremities.

Further, for maximum return on investment in a robot, the work load must be high enough to keep the robot busy. Two or more shifts would put the robot to full use; individual batch sizes should be about 100 parts or more. Parts that require many short welds, curved welds, or welds in a number of different positions usually maximize the cost-effectiveness of robotic welding systems.

Compared to human workers, robots generally perform work of more uniform quality in shorter throughput times. For example, the IRb 60/2 fetches and delivers goods, as well as stacks and positions them in accordance with a predetermined pattern on a standard pallet. The goods can be positioned so that load bearers and transport facilities can be used in the most efficient way. Manual picking operations can be eliminated or reduced to a minimum; the human workers can be put on supervisory tasks or rate work which is not machine-confining.

ASEA's IRb 60/2 robot is geared for heavy-duty operations, typically such tasks as cleaning castings, grinding, deburring, and polishing (Fig. 4-4). All of these operations are monotonous and risky tasks for human workers; cuts, eye injuries, and other dangers have contributed to a high turnover of personnel in such jobs. ASEA has applied adaptive control to deburring castings in the automotive industry, and has developed special deburring sensors mounted in the tool holder in the robot's hand. A constant point of intersection between machined and cast surfaces is achieved, produced with greater accuracy and stamina than can be had with human workers.

Many of the ASEA robots also lend themselves to tasks requiring precise and delicate movements (Fig. 4-5). *Robotics Today* magazine describes a versatile work cell using two synchronized ASEA robots, developed by engineers at ASEA Control (a Swedish-based

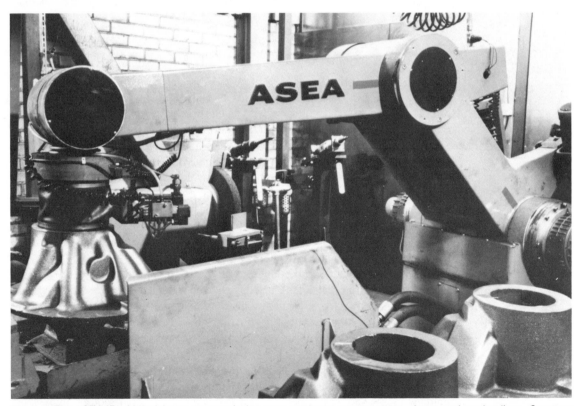

Fig. 4-4. The ASEA IRb 60/2 picks up a transmission case casting to remove gates, risers, and parting lines. Courtesy ASEA Robotics, Inc.

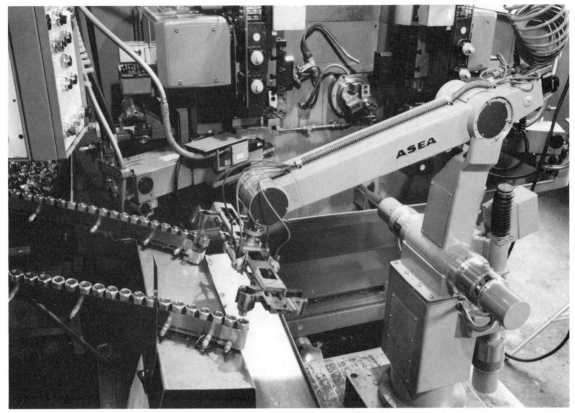

Fig. 4-5. The ASEA IRb 6/2 robot with dual gripper can pick up two brass bushings simultaneously. Courtesy ASEA Robotics, Inc.

unit of the ASEA Group) to build two different sizes of electrical contactors that encompass four different power ratings. After trial runs in an ASEA development laboratory, the work cell was placed in production service in 1982.

The two robots in this cell are six-axis ASEA IRb 6 units. One is floor-mounted; the other hangs directly overhead, suspended from a special support frame. The circular layout of the work cell consists of two general areas, a pickup station and an assembly station. The contactors are fed into position on conveyor belts; vibratory feeders supply smaller components such as springs, clamps, and locking pins. While one robot gathers components from the feeder stations, the other robot performs tasks at the assembly fixtures.

At the beginning of the cycle, one of the robots picks up six contactor components: an armature, coil washer, core, coil, contact guide, and frame. (The gripper used consists of a pneumatically driven turret with six mounting plates for holding six sets of interchangeable robot fingers.) The contactor is put together as a series of subassemblies, which are then combined to form the finished unit. Components are checked using proximity sensors, and completed contactors are then routed by conveyor belt to finishing and testing operations.

The robotic installation is successful. Robotic assembly has eliminated the possibility of common assembly errors (such as using a coil with the wrong voltage rating), and the finished contactors are tested by computer. The payback of the installation is estimated to have occurred in 2.5 years, with the system operating at over 80 percent uptime.

Users and Operations

Teaching ASEA robots is relatively simple. For example, the IRb 90S/2 spot welder is programmed by means of an operator-control system dialogue on a portable programming unit. The control system poses questions in plain language (in the operator's choice of English, French, German, or Swedish!) on an alphanumeric display; the operator/programmer pushes buttons to supply the appropriate answer.

Manual transposition of the robot arm during programming is determined by a proportional-speed joystick on the programming unit. The robot can be ordered to move according to rectangular robot coordinates or wrist-oriented rectangular directions. Programmed positions can be adjusted even during execution, and up to nine different tool center points can be programmed.

ASEA's adaptive control, using point and curve adaptivity together, makes it possible for robots to make program adjustments automatically. Rough programs are prepared using the conventional teaching method; they are adapted automatically in the robot's memory according to the actual conditions and positions encountered. Automatic adjustment reduces programming time and allows the operator to use essentially the same program for pieces having similar—but not identical—shapes. This capability is especially important for arc welding small batches or parts, or for using parts with large production tolerances.

Summary

Name:	ASEA
Manufacturer:	ASEA Robotics, Inc. 16250 West Glandale Drive New Berlin, WI 53151 (414) 785-3400
Physical Description:	Vertical and horizontal arms on pedestal base, with articulated and rotary arm motions; sweep varies from 260 degrees to 340 degrees, depending on model. Five to six degrees of freedom, exclusive of grip function. Handling capacity varies from 13.2 to 132 pounds, depending on model. Electric drive; control system built from plug-in units.
Primary Use:	Machine tending, injection molding, castings cleaning, parts assembly, grinding, deburring, polishing, trimming,

sprue cutting, piercing, hot embossing, spot and arc welding.

Significant Sales Point: ASEA robots are adaptable specialists that can perform diverse jobs in difficult or hazardous environments.

AUTOMATIX AID 800

Automatix, a relatively new, publicly held robotic systems company located in Billerica, Massachusettes, develops and produces modular turn-key robotic and artificial vision systems for a wide variety of industrial uses. One of their best-known products, designed to increase welding productivity and consistency, is the Robovision II Programmable Arc Welding System—also called the Automatix AID 800 robot.

Robot Description

The 770-pound Robovision II Programmable Arc Welding System (AID 800) shown in Fig. 4-6 is shaped like a tall fire hydrant. Extending from its top is an articulated arm assembly consisting of a vertical arm supporting a horizontal arm, with a

Fig. 4-6. An operator programs the Automatix AID 800, a robotic arc welding system designed to increase welding productivity and quality. Courtesy Automatix, The Robotic Systems Company.

wrist/manipulator at the end. The arm/wrist assembly is fast and flexible; the arm bends at 120 degrees per second, and the wrist can twist at 180 degrees per second. Standard accessories are the 880-pound A132 controller, plus the welding equipment.

The wrist/manipulator is all-electric, with five degrees of freedom and a standard torch-mounting configuration. The robot can use a variety of welding equipment; the manufacturer offers a number of standard torches. Both the GMAW and FCAW processes currently are supported. For automatic operation the AID 800 uses a Lincoln LN-9 wire feeder that handles a wide range of wire sizes and features solid-state regulation of line voltage, load voltage, and speed. The welding power supply is a 600-amp Lincoln DC 600.

The AID 800 is served by any of three models of positioners. The AP1-2500 parts presenter is a single-axis turntable with a payload capacity of 2500 pounds; Model AP2-500 has two axes (rotation and tilt) and is rated at 500 pounds. Figure 4-6 shows two AP2-500s mounted atop an AP1-2500 to form the Model AP5-500 five-axis positioner.

The A132 control processor contains a computer-based welding control system and facilities for operating, programming, teaching, and diagnostic display of the system state at any time. The controller is capable of automatic on-line adjustment of welding voltage, torch speed, and wire feed during programming or operation. The robot uses the standard RS-232C serial port for communication and has 16 opto-isolated I/O lines; Automatix plans to add CAD/CAM links in the near future.

The standard memory size for the AID 800 is 1000 program steps or instructions, with optional expansion capability in 1000-step increments. Points or paths may be corrected, added, or erased at any time by manually selecting the appropriate instruction and modifying the data. Many separate part programs can be stored via cartridge tape; stored programs can be selected on command so that different parts may be intermixed and welded with no time lost for reteaching.

Robots depend on computer languages. The Automatix proprietary language for the AID 800 is called RAIL, a conversational control language designed for nonprogrammers. RAIL provides for inter-computer communication, user prompting on the video screen, built-in diagnostics, input of welding offsets, calibration and subroutine calls, and easy interface to positioners, CAD/CAM links, and other robots.

Industries and Applications

The AID 800 is targeted for use in arc welding industries where continuous parts welding is the norm, such as in automobile and machinery manufacture. (Figure 4-7 shows an AID 800 in action.) The main function of the robot is to increase productivity and to ensure consistently high-quality welds. For example, the Automatix computer-controlled rotating positioners with multiple fixtures allow an operator to load and unload parts on one positioner while the robot is welding at the other; as much as a threefold increase in effective arc-on time can be realized.

At the Grand Rapids, Michigan, plant of Steelcase, the world's largest manufacturer of office furniture, the first robot installed (1980) was an Automatix unit that was put to work welding the wire frames for the quarter of a million stack chairs the company produces each year. That first robot changed a four- or five-person operation into a one-person operation and paid for itself in a year. A second Automatix was installed in a Steelcase-designed work cell to weld caster bushings on the base arms of office chairs.

40

Fig. 4-7. The Automatix AID 800 programmable arc welding robot in action. Courtesy Automatix, The Robotic Systems Company.

Labor cost savings were the most obvious benefit for Steelcase, although the company found it was using considerably less welding rod as well. The best thing about the robots, however, could be the quality of work. With some customization (metal inert gas welding units are used to minimize slag), the Automatix robots quickly put the right size weld in the right place at the right time with predicatable accuracy and consistency; hand welding is only as consistent as the human doing the welding.

Steelcase, which employs 6000 persons at its Grand Rapids facilities, has a policy of not laying off workers displaced by automation. When Steelcase purchased its first Automatix robot, the company sent two specialists to train the welders in the operation and maintenance of the robot. Now the robots do the actual welding, but the displaced welders work with them as a team and maintain the welding units.

Users and Operations

The user can easily program many different welding applications into the Robovision II system through a hand held control device called the Interactive Command Module. At each step, the operator is prompted to input a command in the menu-driven software. The robot can be commanded to move in straight-line (cartesian) motion, in robot coordinates (joint angle), or in tool coordinates (along the tool axis). Welding parameters such as voltage, weld speed, and wire feed speed are specified. Once taught, the robot will then automatically and continuously weld along the programmed path.

Summary

Name:	Automatix AID 800
Manufacturer:	Automatix 1000 Tech Park Drive Billerica, MA 01821 (617) 667-7900
Physical Description:	Two-section articulated movable arm and wrist/manipulator attached to a vertical base, plus controller. Robot weight is 770 pounds; controller weight 880 pounds.
Primary Use:	Provides a uniform, repeatable weld that minimizes rejection rate in any industry that performs production arc welding.
Significant Sales Point:	Increases welding productivity and consistency through rotary parts supply and positioning, allowing high arc-on times to be achieved.

CYRO

The technological base from which the Cyro series of robots evolved began to be developed in 1969 by a company known as TekTran, a joint venture of Air Products and Chemicals, Inc., and North American Rockwell, formed to transfer the welding and testing techniques developed in the Apollo space program to a commercial enterprise. In 1973 Air Products and Chemicals purchased the joint venture as a wholly owned subsidiary and, in 1975, moved it to Lancaster, Ohio, and joined forces with another subsidiary, The Arcair Company.

During the 10-year history of TekTran, many firsts in welding technology and robotics were recorded. In 1972, for example, a 40-foot robot was developed for Babcock and Wilcox; this development led to the delivery of TekTran's first welding robots in 1974. In the succeeding years over 50 computerized robots were developed, among them 18 Cyro robots that were the prototypes of those now being manufactured.

Advanced Robotics Corporation, formed in 1979, purchased the assets of the welding products division of Air Products and Chemicals, hired 23 of its key people to help set up the new corporation, and moved operations to the Newark Industrial Park near Columbus, Ohio. The company then set out to build a standardized arc welding robot from the technology that had been developed in more than 10 years of research and design. The result was the Cyro welding robot.

Robot Description

Cyro robots have one horizontal arm attached to a vertical base and a work platform. They all are designed for arc welding, specifically to control the variables of arc welding in a production environment. One of the primary ways it does this is a system called *adaptive control*, which acts to correct the weld path and process parameters in real time.

The Cyro 750. This machine (Fig. 4-8) is an all-electric, 5-axis rectilinear robot that can handle a variety of jobs within its 3/4 meter square work envelope. It is capable of executing the GMAW, GTAW, FCAW, PAW, and SAW weld processes, with exceptional repeatability to maintain program accuracy. The Cyro 750 is equipped with user-friendly software, and it can be taught via a teaching pendant, numerical control through a terminal, or off-line programming.

This robot weighs in at 5740 pounds, with the control apparatus adding another 800 pounds; it requires 27.5 square feet of floor space for the robot base and 6-8 square feet for the controller. Other features include an extendable horizontal axis and a coordinated robotic positioner control. Control hardware features include 64K random access memory and permanent program storage on tape cassette.

The Cyro 820. Described by the manufacturer as one of the easiest-to-use robots available today, the 660-pound Cyro 820 (Fig. 4-9) is a compact, articulated robot suited to limited floor space. (The robot base is only 18 inches square, while the control cabinet requires a space 31.5 × 33.5 inches.) Its features include 5-axis interpolation, repeatability to ±0.008 inches, increased memory size, decreased programming time, and flexible

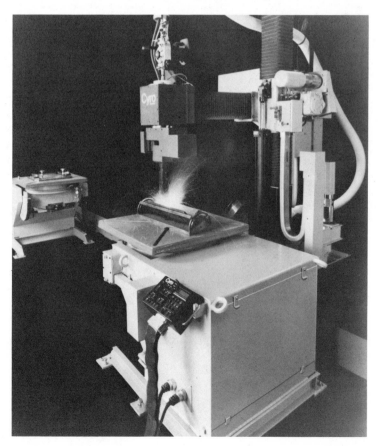

Fig. 4-8. The Cyro 750 robot, designed specifically for high-speed welding, provides a high degree of torch accuracy. Courtesy Advanced Robotics Corporation.

user-friendly software. Programmed through a teaching pendant, it is capable of continuous-path or point-to-point operation; in point-to-point mode it can memorize up to 1000 points per program. Up to eight programs can be stored on cassette tape.

The Cyro 2000. This machine (Fig. 4-10) is a horizontally expandable rectilinear robot suited for seam tracking very long work pieces or traveling to multiple work cells on either side of its horizontal axis. The working volume is a 2 × 2 × 2 meter cube, and its five axes offer a repeatability of ±0.016 inch—unusual for a robot of its size.

Advanced Robotics designs robotics positioners with coordinated motions that manipulate the work piece while the robot is welding, using gravity to keep the piece in optimum position. Used with a tailstock, the positioner turns into a welding lathe that provides rotation for continuous or fixed-sequence welding. The positioners are configured to the particular work cell for two-station welding; while a Cyro is welding at one station, loading or unloading can be done at the other. Arc-on time increases significantly, and waiting time is thereby reduced.

Industries and Applications

Any production-line industry that requires arc welding is especially well-suited to robotic automation. As with other welding robots described in this chapter, the Cyro machines have found wide use in the automotive industry. Howard C. Tuttle, writing

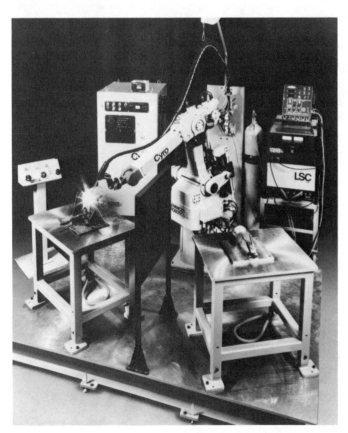

Fig. 4-9. The Cyro 820 welding robot is geared for use on the production line, in the small parts fabrication plant, and in the job shop. Courtesy Advanced Robotics Corporation.

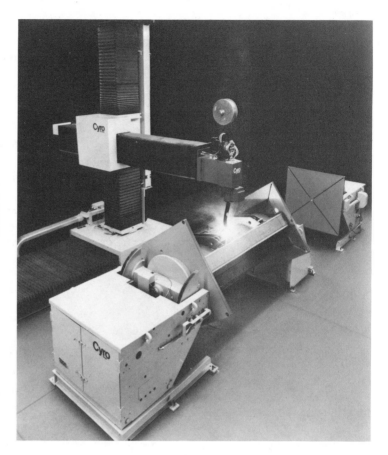

Fig. 4-10. The Cyro 2000 is a five-axis robot powered by electric servomotors. It is built for off-line programming and is designed specifically for welding and seam tracking. Courtesy Advanced Robotics Corporation.

in *Production* magazine, describes a $2 million, 18-robot facility (said to be the first multi-robot arc welding line in the world) that has been in operation since 1979 at the Barrie, Ontario, plant of Hayes-Dana's axle division.

Operating under computer control, the 18 Cyro robots weld six families of 33 different axle housings. This high-volume facility rapidly switches from one housing design to another, with end-to-end dimensions ranging from 73 to 84 inches. Machined wheel spindles and brake flanges are welded at the ends of stamped metal channel arms extending from the bowl-shaped differential housing in the center of the axle. The channels themselves are formed by welding two channel halves along longitudinal seams.

Faced with a "down" economy and stiff foreign competition, Peabody Galion—one of the world's largest manufacturers of dump truck bodies and refuse hauling equipment—saw the profitability of its line all but disappear in the early 1980s. In order to survive, the company chose robotics as the means of increasing productivity while reducing manufacturing costs. A Cyro 820 robot, put to work welding a variety of parts, produced dramatic results: welding time was immediately cut in half, significantly reducing costs, and improved weld quality and consistency allowed the company to reduce the weld size and, in some cases, eliminate secondary welds altogether.

Peabody Galion estimates that the arc-on time for their Cyro 820 could go as high as 90 percent, compared to an average of 25 percent for manual welding, which requires more time for set up, changeover, and clean up. The robot provides about 7 1/2 hours of productive welding in an 8-hour day. Pressure is on to keep the robot running all the time; when the robot is running, the company is making profits.

Users and Operations

Advanced Robotics works with users to create welding work cells designed to fit individual process requirements, while offering logical, functional operation. The Cyro robots are easy to run, and Advanced Robotics will train the work cell operators. Set-ups can be altered by pressing buttons on the control panel or keys at a video terminal, changing the number in a part program, and the robot can be programmed by any one of three methods: a teach pendant, numerical control via terminal, or off-line programming. Furthermore, parameters for the various types of welding processes are built in, and many process control functions, such as adjustments for a weld angle change, are made by the computer automatically.

Summary

Name:	Cyro 750, 820, 2000
Manufacturer:	Advanced Robotics Corporation Newark Ohio Industrial Park Building 8, Route 79 Hebron, OH 43025
Physical Description:	Generally a single arm mounted on horizontal base; configuration of base and arm support varies with model. Weight and footprint varies from 660 pounds and 2.25 square feet (Cyro 820) to 6 tons and 54 square feet (Cyro 2000). Both robots have five axes and are servo-powered.
Primary Use:	Cyro robots are designed specifically for arc welding.
Significant Sales Point:	Simplified programming and control of arc welding variables through adaptive control techniques.

HUBOT

Have you ever found yourself wishing you had a loyal, dutiful personal servant who would bring you a cold one when you asked for it, let out the cat and check the back door lock at night, not take vacations, never give you any back talk—and who might even be persuaded to do windows?

If you have, a Carlsbad, California, firm believes you're not alone. Hubotics, Inc., projects 1985 sales of 4800 units for their $3,495 electronic personal robot called "Hubot."

Sales for 1986 are expected to be 9600 Hubots, a 100-percent increase that translates into a sales growth from $10.3 million to $22.9 million in two years—against competition from at least three other manufacturers of personal robots.

Beyond the most obvious market—robotics hobbyists—lies a wider range of applications than might first be expected. Hubotics, a privately funded company, assembles robotics components into a package that's handy around the house, keeping costs low by having subassembly work done off-site. Hubot can perform a variety of tasks that could make it a relatively economical time and labor saver for the elderly or handicapped; since it will come when called, it could be a blessing to the bedridden.

Hubot is delivered with a label on top that tells the new user to peel it off, turn on the key, and type in his or her name. A menu of things the robot can do appears on the screen, and it says its first word: "Hello <name>, my name is Hubot. How may I serve you?"

Robot Description

Looking like a cross between a juke box and a fire hydrant, with a computer terminal for a head, the rectangular Hubot stands 3 feet 8 inches tall, weighs 65 pounds, and wanders through rooms on a set of lawnmower wheels. An integral part of Hubot is an 8-bit, CP/M-based microcomputer offering 25-line by 80-column video display, ports to connect a modem and the optional $300 printer, 128K random access memory (RAM), and a single disk drive, bringing the total on- and off-line storage available to 512K. (A second disk drive is a $395 option.) Hubot's computer supports programming in both BASIC and FORTRAN.

Additional standard equipment supplied with the robot includes a 12-inch black-and-white television set, AM-FM stereo radio, a cassette tape deck, an ATARI 2600 video game cassette receptacle, a 1200-word voice synthesizer that responds to human requests, and digital time and temperature readouts, all powered by a rechargeable battery. Hubot avoids crashing into things by means of an Obstacle Sensor Processor, and its path can be plotted and memorized with joysticks.

A number of extra-cost options help customize Hubot for individual users. For an extra $150, the owner can get an automatic telephone dialer that responds to a voice command consisting of the name of the desired party. A Smart Servant option allows the robot to control several household appliances; the Sentry package enables it to turn lights on and off, control a thermostat, detect fire or intruders, and automatically dial for help. With an automatic recharger option, Hubot can recharge its own batteries. A vacuum attachment ($300) lets it sweep floors, and an articulated arm ($700) can pick up objects. All the additional equipment can bring the robot's price to $6000.

Industries and Applications

Imagination is one of the few practical limits on household jobs the user can find for Hubot to perform. It can be used like any other microcomputer to run the vast library of CP/M programs or play computer games, as an entertainment device, or as a personal servant—an appliance that can run other appliances.

For example, a user can ask Hubot to awaken him at a specified time each morning; the robot will start the coffee, bring a can of juice, wake others in the house, display an appointment calendar for the day, and supply a printout—all by voice command. It will vacuum the floors, control light and heat, notify the police if an intruder enters, and dial the fire department if it detects smoke.

Hubot will even meet guests at the door with a friendly greeting, by name if so programmed. It says, "Hello. Please come in and sit down." The robot will bring them something to eat and drink and ask, "What kind of music would you prefer to hear, jazz or classical?" It then plays whatever kind of music is requested.

Hubot can be used to assist the elderly or the handicapped, performing household chores of which the user is no longer capable. It can fetch food or drink and, if it were asked to "call Dr. Jones," it would do so with only the voice command for instruction.

Although targeted for the home, Hubots also are versatile enough to find their way into nursing homes, retirement centers, schools, hospitals, and rehabilitation facilities. In addition, the possibilities for promotional use are virtually endless. Hubots can be used by theaters, shopping malls, department stores, amusement parks, arcades, trade shows, and conventions. (See also the sections on Robbie and the promotional offerings from The Robot Factory later in this chapter.)

Users and Operations

Hubot is easy to operate. It comes with mnemonic-based, user-friendly software and a comprehensive, clearly written instruction manual complete with drawings that illustrate how each device works. Many of the functions, such as those for the Sentry and Smart Servant packages, are preprogrammed, needing only the appropriate data.

As noted earlier, motions can be programmed via two joysticks. The user pilots Hubot through the motions of a particular task, a map of which is registered in the on-board computer. For example, you can map out the path the robot is to take to vacuum a carpet. Once stored in memory, the motions will be repeated continuously until superseded by another command.

Summary

Name:	Hubot
Manufacturer:	Hubotics, Inc. 6352 Corte del Abeto Carlsbad, CA 92008 (619) 438-9028
Physical Description:	Rectangular body; head consists of terminal and screen for 8-bit, CP/M-based microcomputer. Height is 3 feet 8 inches; weight is 65 pounds. A range of hardware and software options is available.
Primary Use:	Household management appliance, 8-bit microcomputer, burglar alarm, smoke detector, entertainment device, aid to the

	handicapped, promotional and publicity robot.
Significant Sales Point:	Hubot is a robot servant that greets people, makes phone calls, does housework, fetches food and drink, and detects and acts upon emergency situations in the home.

INTELLEDEX 605

In 1982 Stan Mintz, then an employee of Hewlett-Packard, was searching for specialized capable of installing integrated circuits—silicon microchips—in the circuit boards of pocket computers and calculators. Finding none, Mintz and five other workers left Hewlett-Packard to form their own development company. Its goal: to produce a flexible new robot geared specifically for high-technology electronics assembly work.

Intelledex Corporation of Corvallis, Oregon, was organized with a loan of $150,000 from Apple computer inventor Steve Wozniak, according to a June, 1983, *Robot/X News* article titled "Intelledex Brings Assembly Robots Out of the Woods." Soon after, some $7 million in venture capital flowed into the new company. Today Intelledex continues to be a pioneer in a class of robots called *light assembly robots,* which are designed not for lifting heavy loads, but for tasks requiring dexterity and high precision. These robots are projected to leap from their current 3 percent share of the total robotics market to an 86 percent share by 1986.

The Intelledex Model 605 light assembly robot made its debut in April, 1984. This $48,000 machine emphasizes the flexibility required of industrial robots used in electronics manufacture. Since a great many applications require special tools, adapters, and computer interfaces, the Model 605 was designed as an integrated, maximally adaptable hardware-software system.

Robot Description

The six-axis Intelledex Model 605, shown in Fig. 4-11, consists of a vertical base assembly to which is attached a series of pivoted/counterweighted or axially rotating arms, the last (called the *end effector*) incorporating a stepper motor and a pneumatic pincer mechanism, as well as the optional vision system. On a work table 3 feet square the robot arm assembly can describe nearly a 4-foot circle, emulating human arm and wrist movements. If a sequence of programmed motions causes the upper arm to form an obtuse angle with the table (beyond 115 degrees), the arm assembly must curl; the rear portion of the work circle shrinks only slightly, however, to a radius of approximately 21 inches.

The 605 controller, designed by Intelledex around the widely used Intel 8086 microprocessor, includes an 8087 arithmetic processor and an expandable memory system with 128K bytes of regular random access memory, 2K complementary-MOS RAM, and 8K of EPROM (erasable programmable read-only memory). It accepts input from optical sensors that detect part misfeed, from force sensors that detect the presence

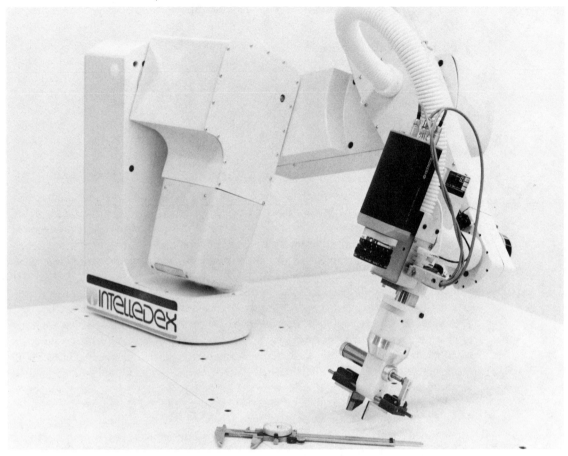

Fig. 4-11. The six-axis Intelledex Model 605 can be used to install integrated circuit chips or inspect component placement on printed circuit boards, and is sufficiently sensitive that it can be used to install the heads on hard disk drives. Courtesy Intelledex, Inc.

of an object in the end effector and sense excessive pressure, and from a bar code reader that picks up identification information on assembly components. To ensure safety, the controller also monitors input from pressure-sensitive floor mats and a light curtain composed of photocells surrounding the work areas; work cannot begin until the area is clear.

Software is written in Robot BASIC, an enhanced version of the popular Microsoft BASIC that features nearly 150 additional, robot-specific program commands. This specialized, high-level language (which runs easily on most personal computers) allows the Model 605 to be programmed like a computer, and to compensate for mismatched accuracies or parts placements. The robot can adapt itself to different, multiple work-plane orientations that can be defined through software, and can work at two tables that have different positions relative to the robot arm. It even can continue to work while a new program is being written for it on a remote terminal or personal computer.

To control the robot manually for positioning and path-point entry, as well as the functions of speed and tool operation, a *teach pendant* containing a joystick and 18 switches is provided.

The Model 605's optional vision system is capable of recognizing up to 100 different electronic parts. In designing the system Intelledex departed from industry practice by choosing a 6-bit video converter over the customary 8-bit devices. The company thus was able to simplify its software, so that users can operate the vision system without having to learn assembly language or other specialized code, and to cut drastically the system cost. The Model 605 vision system sells for $12,000 to $16,000; were it based on an 8-bit converter, the company says, the price would be in the $50,000 range like comparable systems from other manufacturers.

Industries and Applications

Designed specifically for precision assembly operations, the Intelledex Model 605 can and does find wide use in the electronics and computer manufacturing industries. The robot can be used for inspection of component placement on circuit boards, orientation and alignment of assembly components, and real-time monitoring of the component inventory. The robot is used to install silicon chips in pocket computers. It can check polarity markings on capacitors. The 605 may be used where components with wire leads complicate the assembly process.

Most assembly operations involve a continuous flow of components to an area called a *work cell*. The parts have to be aligned and, when the work is finished, the product must be removed. The Intelledex robot lends itself well to tasks that require little preparation of assembly pieces, or it can be programmed to perform many of these preparatory steps itself. Circuit boards, for example, must be brought to the work cell and aligned in a fixture that utilizes circuit board tooling holes. The robot can be used to unload the boards from racks, align them on the tooling fixtures, and then perform whatever work is necessary.

Manufacturers of tape cartridge drives, such as Data Electronics, Inc., of San Diego, California, have found the 605 robot/computer system useful in their work. At Data Electronics six types of odd-shaped components are loaded into circuit boards that already have been stacked with standard components; the boards are supplied to the cell in racks, directly from automatic insertion machinery. The robot selects eight of the components and inserts them in the board, which then goes to a final inspection station and a flow-soldering machine.

Intelledex Model 605 robots also are used to place head assemblies in Winchester hard disk drives; the process is discussed in detail in the next section.

Users and Operations

The Intelledex 605 robot can be operated easily by a user who is accustomed to electronics assembly work but who is relatively unsophisticated in the use of computers. Because the vision and software are integrated in the robot, assemblers who formerly installed parts in circuit boards manually can be taught in a short time—usually a matter of days—to instruct the robot to do so.

Assembly of Winchester disk drives provides a fairly typical example of the kind of work the Intelledex does best, as well as how the robot and its human operators work together as a team. Let's follow it through, step by step.

The Model 605 is positioned centrally on a work surface called a *shuttle table*, with operators at each end supplying disk drive parts to the robot. To begin a cycle an operator places a disk drive in a movable support/clamping device (called a *nest*) on the table, clamps it, and then places a head assembly on a pair of positioning pins. He or she then slides the nest into the loading position. A switch signals the safety computer that the disk drive is in the correct place, where upon the computer signals the nest to lock.

After inserting a comb-like piece to separate the heads, the robot grasps the head assembly in the gripper and, acting on information from the vision system telling it the location of a screw hole in the disk assembly, positions the head assembly, drives home the screw, pulls the comb from between the heads, drops it in a box, and stops. The computer tells the robot to remove the locking pin, and the completed head-disk assembly is slid from the loading position. The robot's cycle time for this operation, repeated for each drive, is 60 to 90 seconds.

Intelledex has an excellent robotics customer training program, which is examined in greater details in Chapter 6 of this book. The program is divided into several modules. Module 1 provides an overview of Intelledex robotics. Topics covered include information about the company, a tour of the facilities, and a general introduction to the operation of the robot. Trainees then learn about the parts of the robot, modes of operation, vision, system software, tool design, and safety. Module 2 is completely devoted to hands-on robot training, and Module 3 teaches the participant to troubleshoot and service the robot and vision system.

Summary

Name:	Intelledex Model 605
Manufacturer:	Intelledex Corporation 33840 Eastgate Circle Corvallis, OR 97333 (503) 758-4700
Physical Description:	Vertical stand with multi-jointed arm ending in wrist with pneumatic grippers. Six axes, with repeatability ± 0.001 inch. Load capacity is 5 pounds at maximum speed, or 12 pounds at reduced speed.
Primary Use:	Assembling computer components; other electronics assembly work.
Significant Sales Point:	Ease of operation; little user training required.

JPL ROVER

The Jet Propulsion Laboratory (JPL) at the California Institute of Technology in Pasadena has maintained a robotics and machine intelligence program since the early 1970s. The program has since grown to include research on robots to be used in space exploration and industrialization, as well as robotics applications on Earth.

The primary objective of JPL's robotics research is to combine new and existing technologies to create robots capable of performing tasks and making decisions with a minimum of human direction. These robots and independent machines will serve in a broad spectrum of industrial and scientific applications.

In 1977 the JPL robot research program attained its first major goal, demonstrating that their robotic hand-eye system could be combined with a self-propelled vehicle for planetary exploration. Future mobile robots will use on-board computers to store data obtained by imaging and sampling instruments, relaying the information back to Earth or to orbiting spacecraft.

One of the main requirements JPL has established for its robots is *system autonomy*, a characteristic necessary for space exploration. A remote machine with independent decision-making capabilities can perform tasks without step-by-step instructions from its human supervisors. The robots contain compact memories and processors that provide them with portable machine intelligence. They can sense their environments, plan and execute actions, and perform manipulations requiring a degree of dexterity normally reserved for humans. Many JPL robots are controlled via *teleoperation,* a process developed in the 1960s to augment and extend human sight, touch, and cognitive abilities to remote locations.

For operations in space JPL designs its robots along strictly functional lines; they may take the form of Earth-orbiting telescopes, planetary fly-by vehicles, or spacecraft. Some JPL robots are stationary landers equipped with manipulators to handle soil and rocks, while others are configured primarily for mobility. The JPL robot discussed here, called the Rover, is a research tool designed to study the problems of combining visual and manipulatory systems to be used on planetary exploration vehicles and on craft such as the space shuttles. But it is more than just a robot, however. The Rover is a forerunner of a vast diverse array of extraterrestrial robotics applications.

Robot Descriptions

About the size of a large office desk and roughly automotive in form, the JPL research Rover (or *Mars Rover*) shown in Fig. 4-12 measures 59 inches long by 51 inches wide and weighs approximately 700 pounds. Its instrument payload is 220 pounds. The vehicle is equipped with a single manipulator arm, twin camera pylons, and an antenna mast. Future versions will be equipped with additional manipulators and drills to retrieve rock and soil samples, and will power all on-board systems with a 200-watt integrated radioactive thermal generator (RTG).

The Rover travels by means of a mobility system consisting of wheel and track assemblies (called *loopwheels*) similar to tank treads, located at the ends of its four articulated legs. A knee-action suspension with interconnected "knee" and "thigh" joints allows each loopwheel foot to adjust independently to varying terrain, maintaining the Rover's box-like chassis in approximately a level position. Moving at a speed slightly over 3 feet per minute, it can negotiate 24-inch obstacles or depressions (like boulders or craters) and can climb or descend slopes as steep as 30 degrees.

The prototype guidance system under development by JPL consists of two television cameras mounted on pylons above the chassis and tied into an on-board computer system. These stereo cameras serve as the robot's vision. A chassis-mounted laser range

Fig. 4-12. The Jet Propulsion Laboratory's intelligent, desk-size Mars Rover robot is designed to traverse at least 100 kilometers of the red planet's surface during a one-Martian-year mission, deploying scientific packages along the route, studying the surrounding terrain, remembering obstacles, and avoiding hazards—independent of detailed instructions from Earth. Courtesy of Jet Propulsion Laboratory, National Aeronautics and Space Administration.

finder can measure the distance between the Rover and any obstacles. The robot's computer brain correlates and interprets the video and range finder data, charts the best course to a given destination, and then instructs the vehicle to move along the specified path.

The Rover's computer uses the vision system to determine the relationship of the manipulator arm to any objects it will handle. (JPL is developing proximity sensors that will enable the robot to "feel" the distance between manipulator and object.) The system is programmed for a degree of manual finesse sufficient to allow the robot to pick up and hold a cylindrical pin in its two-fingered "hand," find a round hole in a metal block, and, using a force sensor, slip the pin into the hole without binding. The particular way these functions are integrated enables the robot to work with moving objects. (This talent may be less important to the Rover than to future space-borne robots that might have to track objects such as satellites or orbiting bombs and pluck them from the sky.)

Industries and Applications

The best locations for autonomous robot vehicles obviously will be in space or on planets other than Earth; they will be used for a variety of scientific objectives. In the course of future missions, vehicles such as the Rover will be constantly moving to new positions on a planetary surface. It will encounter many different geographic features, make a daily recording of its surroundings and relay what it perceives back to Earth. Operators will issue general commands; the robot will then proceed autonomously throughout the day, performing tasks and scientific investigation with a minimum of instruction from Earth.

Space industrialization will depend heavily on robotics and automation technology. Industrial activities in space will include the construction of large antenna arrays, space stations, and solar power systems. These will be built largely by robotic cranes, manipulators, and teleoperators. Manufacturing and processing stations in space will be operated and maintained through automated instrumentation and control systems.

According to Jet Propulsion Laboratory releases, future automated space systems now under development or construction at JPL include the following:

- Orbiting manufacturing modules to produce biological, metal, and liquid products for research and commercial use.
- Systems to provide on-board health care to passengers and crew of manned missions.
- Lunar rovers designed to collect and stockpile materials for the construction of lunar bases.
- Lunar base systems to perform mining, processing, and plant operations.
- Remotely controlled or manned carriers to deliver nuclear waste from Earth for disposal in deep space.

A recent JPL study estimates that NASA could save as much as $1.5 billion through A.D. 2000 via serious implementation of machine intelligence. Modern computer systems are capable of extracting relevant data in user-compatible form either on the ground or aboard a spacecraft. The desired output might be a direct graphic display of snow cover, crop health, global albedo (the fraction of incident light or electromagnetic radiation reflected by a planetary surface), mineral resources, hydrologic cycle, or storm system development. Machine intelligence can analyze and reduce this data, presenting technological results in readable, convenient form.

Further, the quantity of data made available by NASA missions is huge: the Viking missions alone returned data that filled more than 75,000 reels of tape, and it has been estimated that every 2 years in this decade NASA computers will process and store the equivalent of the entire written content of the Library of Congress. Advanced robotics and machine intelligence systems on spacecraft can and probably will take over much of the data processing and information sorting clerical activities that currently are performed on the ground.

Summary

Name: JPL Rover

Manufacturer: Jet Propulsion Laboratory
California Institute of Technology
4800 Oak Grove Drive
Pasadena, CA 91103
(818) 354-4321

Physical Description:	Vehicle, 59 inches long and 51 inches wide, shaped like a small automobile. Weight is 700 pounds. Equipped with one manipulator arm, four tracked mobility modules at ends of articulated legs, and a variety of sensory apparatus. Can negotiate 30-degree slopes and 24-inch obstacles at maximum speed of 1.0 meters per minute.
Primary Use:	Extraterrestrial exploration, relaying scientific and technical data back to Earth for analysis and interpretation.
Significant "Sales" Point:	The JPL Rover requires very little human intervention to perform its tasks.

KUKA

KUKA industrial robots are manufactured by Expert Automation, Inc., of Sterling Heights, Michigan. Since 1933 this company has been designing and building special welding and assembly lines, as well as a variety of patented conveyor, transfer, and index table drives.

Today Expert Automation, now a subsidiary of KUKA Welding Systems and Robot Corporation of Augsburg, Federal Republic of Germany, supplies turn-key welding and automation systems to industry. The combination of Expert Automation's extensive work in welding and KUKA's well-known expertise in robotics make Expert/KUKA a strong competitor in the field of automated welding systems.

Expert Automation manufactures KUKA industrial robots in a 57,000-square foot factory that also houses one of the best robot demonstration and training facilities in the nation. The expert Automation plant can handle parts ranging in size from spark plugs to large body-side members. The assembly area permits welding machines up to 240 feet long to be assembled and run off. A combination 6-station rotary plus 11-station inline transfer welding line can join five separate components to make complete automobile radiator support assemblies at a production rate of 600 per hour. A 3-station welding line 60 feet long and 20 feet high is capable of spot welding 60 van body side assemblies per hour. All of the assembly and demonstration lines are fully automated and computer-controlled.

Robot Description

In general, the five KUKA models discussed here consist of horizontal arms attached to vertical arms, which in turn are mounted atop rotary bases. The bases may be floor- or portal-mounted, depending on the needs of the particular application. Each is designed for a specific welding or assembly job, although they can be adapted to a variety of uses. All KUKA robots are subjected to a 100-hour predelivery service test.

The IR 662/80 Robot. The IR 622/80 is a six-axis, floor-mounted robot specializing in assembly-line automotive welding jobs. It uses electromechanical drive and computer control to achieve high speed motion sequences with a repeatability of ± 0.040 inch. The IR 622/80 can handle 200 pound loads, can operate from floor level to heights of 10 feet, and has a travel of 67 inches. The main axis sweep is 320 degrees.

The IR 160/45 Robot. This model is a versatile, general-purpose, electrically operated six-axis robot that can be mounted on the floor or overhead. It features two overlapping 250-degree radial sweeps that permit operations in front of or behind the robot.

The KUKA IR 200 Robot. The KUKA IR 200 is a five- to seven-axis gantry-type robot primarily designed for spot welding operations in which rocker guns or welding cylinders are used against backing electrodes, generating a welding force up to 1100 pounds. This model occupies no floor space within its working area, and can travel within a projected arc of approximately 8 1/2 × 20 feet at speeds ranging from 3.9 to 11.2 feet per second. Repeatability is ± 0.040 inch.

The IR 600 Robot. This system, successor to Expert Automation's IR 601/60, is a six-axis robot suitable for welding and assembly operations; although more robust than its predecessor, it features simpler structure, a more favorable price, and a 10- to 15-year service life. The system contains a portal-mounted, continuous-path Ir 160/60 robot, which is suitable for fitting automotive drive shafts.

The IR 260/500 Robot. The IR 260/500 robot can drill holes in curved metal sheets. The robot is fitted with two drilling units which it uses alternately, keeping them constantly perpendicular to the surface of the work piece, no matter how complex the shape of the metal sheet might be. One application has been the welding of truck panels, working alternately on rear panels of different sizes; the appropriate welding program for each case is initiated automatically.

Expert Automation/KUKA robots employ a continuous-path control system that makes them highly adaptable to frequent model changes and mixed-batch production. Teaching is accomplished through a light, hand-held programming unit which has a "drawing" function to input movement paths, stopping points, and speeds into the system; a menu-driven master control program further simplifies the robot's programming and operation. Along with continuous-path programming, a further refinement which contributes to virtually flawless weld seams is a *waving sensor*. In this feedback system, an electrical signal taken from the arc is evaluated as a correction and control signal for detecting the center of the seam.

Industries and Applications

Although they have been put to use in applications as diverse as installing car doors and wheels, cleaning castings, charging annealing furnaces, and palletizing bundles of glass tubing, Expert Automation/KUKA robots have found widest application performing welding operations in the automobile industry. Capable of both friction and magnet-arc welding, they are adaptable to almost any job in the industry that uses these processes on an assembly-line basis.

In the friction welding process, both components to be joined are securely clamped. The joint faces are brought into contact by a hydraulic cylinder and then one is rotated against the other to produce heat. Once the materials reach plasticity, they are forced together under an extremely high pressure (called *upset force*) to produce a weld of equal or better strength than the materials being joined. Excess material produced by the process (called *weld flash*) is minimal, and can be removed automatically within the machine. Friction welders can be used to join a wide variety of similar or dissimilar materials; eight different models are available, rated from 1 to 250 tons upset force and capable of handling parts ranging from 0.120 to 8 inches in diameter.

In magnet-arc welding, both parts to be welded are electrified and brought together within the field of a magnetic coil. Separating the parts slightly (approximately 0.080 inch) forms at the point of weld a high-temperature electric arc which is controlled and caused to rotate by the magnetic effect of the coil. When the parts have been heated sufficiently they are pressed together and forged. This process produces a porosity-free weld with fatigue strength properties equal to or greater than the parent materials.

The automobile and truck industries have found it increasingly necessary to automate their use of these welding processes to meet production objectives, which include (according to *Industrial Robot* magazine) retooling production lines for different models or component operations, making better adaptation to capacity changes for increased or decreased production rates, mixing production of different models on the same production line, and realizing greater manufacturing efficiency.

The prerequisites for the effective use of robots in car body manufacture, also outlined by *Industrial Robot,* are the creation of reference points for all models, robot-adapted car body design, assembly-adapted car parts, accurately manufactured stampings, centralized production control, and qualified maintenance personnel. The robots themselves require accurate positioning devices, the proper welding tools, carrying and conveying fixtures, data storage systems, and sequence controls.

The Expert Automation/KUKA machines are flexible robot systems capable of meeting these objectives. Typical parts welded by these robots include axles, drive shafts, exhaust valves, fuel tank filler tubes, and steering column components. The robots in the IR 600 series can respot complete car bodies, reaching even the least accessible locations. They are capable of welding complete side panel assemblies for garbage trucks. In a multi-transfer assembly line, they can completely fabricate 13- through 16-inch wheel assemblies—from welding the disk and rim, through punching the valve stem hole, to applying decorative or identification stampings—at a rate of 900 per hour.

Users and Operations

Expert Automation/KUKA offers a comprehensive, hands-on set of structured robot training programs designed to produce skilled, technically competent personnel who have a thorough understanding of the capabilities, operation, and maintenance of the KUKA line of robots. These courses, which include both theory and practical experience, are designed to be responsive to customer needs, as well as to enable participants to operate the robot, create user programs, and clear simple problems. In-plant training at the customer's facility also can be arranged.

The first course, programming and operations, includes an introduction to the industrial robot, robot construction, drive mechanisms, integrated circuits, and feedback systems. The operations phase covers the actual operation of the robot, programming and application of control programs, locking devices, safety installations, routine maintenance, error tracing, identification of problem cause, and repair. A more advanced course covers electrical maintenance, while a third concentrates on mechanical maintenance for repair personnel.

The training courses generally run one or two weeks. Prerequisites for the beginning courses include familiarity with industrial machinery; background for the advanced electrical maintenance course includes knowledge of digital techniques, interpretation of logic circuits, and knowledge of digital/analog control systems. Expert Automation's training courses are also open to the public at a tuition of approximately $850 per week.

Summary

Name:	KUKA
Manufacturer:	Expert Automation/KUKA Inc. 40657 Mound Road Sterling Heights, MI 48078 (313) 977-0100
Physical Description:	Varies. General configuration is six axes, with horizontal arm extending from vertical arm on floor- or ceiling-mounted base. KUKA IR 200 is five- to seven-axis gantry robot occupying no floor space.
Primary Use:	Automotive industry spot welding applications.
Significant Sales Point:	Automatic welding system capable of friction and magnet-arc welding.

MOBOT

Mobots industrial parts-handling robots (the world's largest) have been in factory service since the Mobot Corporation was founded to supply its customers, according to the company motto, with "Practical Robots for Your Real-World Factory." Mobots are used worldwide to load and unload other devices like machine tools, skids and pallets, conveyors, molding machines, and printed circuit manufacturing machines; other Mobots handle unsupported products, including delicate glassware.

Robots find actual use in industry for three purposes only: parts handling, spray painting, and welding. Work requiring dexterity, vision, or judgement is still performed by humans. This perception of modern industrial reality, noted by Mobot Corporation president Lawrence J. Kamm in a letter to the Securities and Exchange Commission accompanying Mobot's registration documents, led the San Diego-based firm in 1974 to develop affordable, practical robot manufacturing machines.

The Mobot design philosophy differs from more conventional robots, which are called *anthropomorphic robots* because their motion capabilities are patterned after those of the human body. By purposely limiting these capabilities and designing around fully modular structures, Mobots can be priced in a range from $15,000 to $40,000. Prices for anthropomorphic industrial robots, in contrast, can start at $45,000 and climb well into the six-figure range.

Second-generation robots generate straight-line motions directly by means of vector motion modules. (Mobot calls theirs "Vectrons.") Rotary motions, usually required only for orientation changes, also are generated directly. Where first-generation, swinging-arm robots require complex computer-generated and servo-synchronized motions to approximate straight lines, second-generation vector robots such as Mobots need fewer axes to accomplish the same tasks, offering much greater travel distances, simpler software, less electronic hardware, higher reliability, and easier maintenance—in addition to the lower system cost.

The average annual cost of a factory worker to his or her employer is $25,000, which figure includes wages, taxes, and benefits. Average Mobots are said to pay for themselves in a year, even working a single shift. Since a one-year payback is considered a universal rule of thumb for evaluating the purchase of cost-reduction equipment, the Mobot Corporation appears to have made more than good on its promise.

Robot Description

Mobots are essentially point-to-point machine tools used for parts handling. They are designed to load and unload materials, and to transport them long distances (currently 65 feet, with 140 feet under development). A special application, the carousel Mobot, transfers materials from rotary to linear conveyors. Developed for conditions where floor space is crowded, Mobots are built on columns and have most of their mechanism located in the free space above the customer's equipment (Fig. 4-13).

The system resembles a kind of Erector® set, a concept the manufacturer calls

Fig. 4-13. Illustrated here are two Mobots working under the control of a single programmable controller; this installation is part of a system of three similar Mobot systems built for a major electronics company. The grippers (center) are built to a proprietary design and have been masked for this photograph at the purchaser's request. Courtesy Mobot Corporation.

modularity. Mobot provides an inventory of linear- and rotary-motion Vectrons in a wide variety of sizes, distances of travel, position control options, and power options; the columns, couplings, brackets, and other structural members available are similarly interchangeable. Standardized parts mounted on the robot's surface provide easy installation and repair access. (Mobot Corporation boasts that it purposely makes an ugly robot because the exposed "innards" make it easy to repair.) Modularity makes it possible to build a Mobot on-site with precisely the motions required for a particular task (and no more)—and then rebuild it to provide the new set of motions needed for a new task.

Most parts-handling operations require a combination of straight-line and rotary motions; Mobots use linear Vectrons to produce both types. Anthropomorphic robots, on the other hand, simulate linear motion through precisely synchronized joint rotations—a technology that creates jobs for servo engineers and computer programmers, but also creates a kind of technological overkill that is not cost-effective for many tasks that otherwise would lend themselves admirably to roboticization. A two- or three-motion Mobot often can do the job of a five- or six-motion anthropomorphic robot, and do it without the need for computers or servos.

For control purposes Mobots need only standard, commercially available programmable controllers, plus the appropriate sensors and encoders from the wide variety of optical, touch, and proximity types available. No other computer is needed, although host computer systems can communicate with these robots via a standard RS-232C interface. Mobots also have manual operating controls used for set-up, debugging, and maintenance; manual controls operate through the programmable controller, however, to ensure safety. For very simple Mobots, the manufacturer offers a low-cost, non-electronic controller using standard telephone components.

Position and sequence programming of Mobots is accomplished either by point-to-point numerical control entered via the programmable controller keyboard, or by a method called "jog-and-record." The latter method corresponds to the "teach mode" of first-generation, swinging-arm robots and uses a "teach box" controller to enter the commands. The gripper portion of the robot also is programmable, allowing the use of quick-change grippers, adjustable grippers, or automatic gripper changes using a "gripper" gripper. The Mobot System Epsilon automatic machine tool uses portable tool magazines having capacities of up to 100 tools, making batch changeovers extremely easy to program. Mobot programs are stored on cassette tape and can be plugged in as needed.

By the addition of special types of Mobots—the MS/RS models—standard industrial carousels can be upgraded and combined with other devices to form automatic storage and retrieval systems. MS/RS Mobots are modular cartesian-coordinate robots which, like all Mobots, are made of standard components individually configured for each customer, providing the precise functions needed to fulfill the task at minimum cost.

The vertical structure of the robot, based on a heavy-duty square box mast, is available in two configurations: single-column models can lift up to 250 pounds, and double-column carousel Mobots can lift 1000 pounds. The mechanism travels vertically on a roller-bearing carriage running on precision steel rails. Two safety chains operating in tandem (each rated at 8000 pounds) are monitored by a sensor system that checks for slack or broken chain conditions. Lifting gearboxes of carousel Mobots are self-locking, and feature normally-on brakes to prevent the load from falling during power failures.

Carousel Mobots are equally usable with fixed shelves or flow racks, single carousels, multiple carousel arrays (up to 20 feet high and 50 feet wide), single or multiple conveyors, automatic guided vehicles, and towline carts. They can be set up in stationary format to serve a single carousel or rack, or they can move horizontally to serve many devices. Single or multiple payload types are supported, and the robots can manipulate full-or half-width totes. There is no limit to the horizontal travel, number of carousels, or number of conveyors serviced. Two or more Mobots may operate over a common track.

Industries and Applications

Mobots are designed to operate in diverse environments: heat, dust, and abrasive environments; corrosive and solvent vapor areas; explosive atmospheres; ultra-clean rooms; and the general factory environment. Coupled with their highly modular structure and interchangeable parts, this feature makes Mobots adaptable to a wide variety of industrial and scientific applications.

Mobots can be used for any parts-handling job in factories. Mobot Corporation robots handle extremely heavy loads—up to 1000 pounds—and transfer them exceptionally long distances; this tonnage previously would be moved by forklift trucks or operator-driven overhead cranes. These robots provide the muscle in factories where loading and unloading is a major activity, factories which frequently do not employ workers with the skill levels necessary to maintain sophisticated computerized machinery.

Mobot Corporation provides machine tool manufacturers with systems tailored to their products. Mobots are used by material-handling and robot systems contractors who specify and install robots as components of complete material handling systems. Mobots place gates into trim ties and extract finished moldings from injection molding machines in the molded plastics industry, sort computer parts in the computer industry, and transfer boards from stacks to cleaning machines in the printed circuit board industry.

Factories using the Mobots usually engage in multiple machine tool and gauge applications. The Mobot System Beta replaces the machining cell clusters required by first-generation, swinging-arm robots. The Mobot System Epsilon noted earlier is a complete, automatic machine tool capable of changing its own working tools. And, as the supply apparatus for machine tools, Mobots can sense and remove from a pallet or cart the top sheet of a stack of sheet material that is to be fed to an automatic punch or shear.

Mobot System Alpha uses a full bridge Mobot to transfer parts from a fixed-position input tote to a machine tool, and then to a fixed-position output tote. Within the totes, parts are nested in plastic dunnage sheets that also are transferred from input to output. Automobile industries worldwide are among the chief users of Mobots. Totes as large as the 6 × 3 × 3 size can be accepted, and the use of multiple totes enables a Mobot to switch among full and empty input and output totes without interruption.

Current research and development projects under way at Mobot Corporation include welding and spray-painting robots that use the company's modularity and vector motion design philosophies. Mobot can provide configurations for spot and arc welding, but currently does not have electronic control. The spray painting robot, which is to use a customer-provided electronic control system, already has been built. Like other Mobots, the overhead mechanism minimizes floor clutter while allowing the robot to move to all parts of an object being painted. Neither robot, however, is offered commercially at this time.

Users and Operations

Mobots can be understood and maintained by regular factory maintenance people and manufacturing engineers after a single day of training; they are built to be operated by persons with no computer background. More complex robots, on the other hand, can require weeks of training and skilled technicians. The manufacturer uses this capacity for quick, one-day operation and maintenance training as one of the main selling points.

Mobot Corporation offers free training for its customers' employees, with emphasis on using the customer's own robot. When a Mobot is ready for shipment, the company invites the customer to send the employees who will be involved with the robot to Mobot headquarters in San Diego, where the company will spend as long as the customer wishes—a day, several days, even a week—to train all of the operation and maintenance crew. Works-based training permits adjustments to be made or a different robot substituted if unanticipated problems arise, before the machine is shipped and bolted down in the customer's facility. Mobot Corporation also is willing to conduct on-site training, but claims it will resist shipping the robot if a training opportunity is not in some way provided for.

The philosophy behind Mobot's training program is that the customer should know as much about the machine as Mobot does when the robot leaves the factory, and that the operator should be able to repair his or her own robot. As a result, Mobot Corporation claims it almost never gets service calls. By training a customer's operation and maintenance people on the machine they'll actually be using (instead of on a demonstrator), and by making the training session a sort of mini-vacation in San Diego, Mobot strives to make each trainee feel that the robot is his or her machine, and that he or she is going to make it work.

Summary

Name:	Mobot
Manufacturer:	Mobot Corporation 980 Buenos Avenue San Diego, CA 92110 (619) 275-4300
Physical Description:	Varies. Most Mobots travel on overhead rails, like a bridge crane, for unlimited horizontal distances. Carousel Mobots move vertically on a single- or double-column mast.
Primary Use:	Loading, unloading, and handling large, heavy loads. Servicing single or multiple conveyors and carousels.
Significant Sales Point:	Mobots are unsophisticated, easy to use and repair, and occupy little floor space of their own.

PEDSCO

A few years ago, an epidemic of terrorist bombings in New York motivated that city's police department to investigate the use of robots to deactivate bombs and secure areas. Among the robots tested was a two-armed, six-wheeled, shotgun-welding Canadian device manufactured by Pedsco, of Scarborough, Ontario. The Pedsco robot, costing from $20,000 to $50,000, is used by the bomb squads and SWAT teams of several U.S. police departments, and works regularly at nuclear reactors both here and in Canada.

Robot Description

The Pedsco robot resembles a tank-type vacuum cleaner that rolls on six wheels. Attached to the body are two arms with hands that can be operated by remote control. One of the hands has a lifting capacity of 70 pounds, and both are adjustable for "soft touch" or "powerful grip" modes; the mode chosen depends in part on the weight of the object being lifted. The robot also contains a boom that can be detached from the main structure by removing two pins and a multiconnector.

The robot's vision is provided by a television camera, also remote-controllable, that can see through 360 degrees; the manufacturer states that "x-ray vision" is an available option. The camera is removable and can be mounted above the equipment if the operator desires manual operation. The camera also can be placed on other parts of the body if the top area is needed to mount another of the Pedsco robot's primary options—a riot shotgun. In addition to its obvious purpose, this weapon also can be used to detonate an explosive device that fails to respond to less extreme persuasion.

Industries and Applications

The main function of the Pedsco robot is to remove dangerous substances or intruders from its assigned area, warn humans of fire or other dangers, and operate in places too dangerous for humans to enter. Although manufactured primarily as a bomb disposal device, the Pedsco also lends itself to use with radioactive materials, in terrorist or hostage situations, for sentry duty or security service, and in firefighting.

Police in Phoenix, Arizona, have been using a Pedsco robot for explosives disarming and disposal work since 1981. Miami and Detroit also have these robots on their bomb squads, and other jurisdictions (including New York, as noted earlier) have shown an active interest in robotics for such applications.

Bruce Nuclear Reactor, Ontario, was in 1982 the scene of a near-accident that put the Pedsco robot to an extreme test. High levels of radioactivity, said by a Pedsco official to be fatal to unprotected humans within one second, prevented workers from cleaning up spilled atomic fuel; to compound the problem, the fuel had fragmented into very tiny pieces. The Pedsco robot was sent in and, operating by remote control, was able to recover both fuel and the radioactive dust created by the spill.

Users and Operations

The operator of the Pedsco robot can be trained quickly and does not need to be a skilled programmer to put the robot to work. Operation is carried out from a remote position via electrical controls and a television camera. Since the robot does, however, mimic the skill level of its teacher, the programmer does have to have the requisite manual dexterity for the job.

Summary

Name:	Pedsco Robot
Manufacturer:	Pedsco
	12 Principal Road, Unit 2
	Scarborough, Ontario M1R 4Z3
	(416) 755-3852
Physical Description:	Horizontal, cylindrical body rolling on six rubber tires. Two arms with force-adjustable hands. Video camera and remote control capability. Optional riot shotgun.
Primary Use:	Surveillance, bomb defusing and removal, nuclear waste disposal, firefighting, riot control.
Significant Sales Point:	Performs tasks dangerous to humans by remote control.

PROWLER

Robot Defense Systems (RDS), Inc., of Denver, Colorado, designs and manufactures mobile robots for military and security applications. The company's chief product is a six-wheeled, tank-like unmanned vehicle called the Prowler that ranges in price from $250,000 for a demonstration sentry model to $500,000 or more for advanced model capable of fighting other tanks.

Robot Defense Systems, a small company that spun off in May of 1983 from Fared Robot Systems, Inc., was founded by Christy A. Peake (who also had been Fared's founder) to research, design, develop, manufacture, and market a military/security robot—the Prowler—the concept for which had originated a year earlier. RDS "went public" in November, 1983, and introduced the first Prowler in 1984. Its manufacturer foresees the Prowler (the name is an acronym for *Programmable Robot Observer With Logical Enemy Response*) as the basic building block for a group or series of battlefield robots.

Battlefield (military) robotics is currently one of the hottest topics among military strategists and planners in all parts of the world, the theory being that the next war should be fought by robots, not soldiers, in order to save lives. The military also is developing its own robots (see the section on U.S. Navy underwater robots near the end of this chapter), and according to *Design News*, U.S. military spending for robotics leaped from $17.9 million in fiscal 1982 to $27.6 million in fiscal 1983. Robot Defense Systems already has spoken to the U.S. Air Force and the government of Israel about possible sales, even as other new and small robot development firms are trying to penetrate this growing field.

The three production versions of their demonstration Prowler unit RDS anticipates are: a "defensive model" to patrol air bases, nuclear weapons stockpiles, and other high-security areas; a "pointer model" that would be towed into a combat area by another vehicle and then be sent in ahead of any army unit to photograph the terrain with television cameras; and an "offensive model" to destroy enemy tanks and troop carriers, firing missiles and dashing from one hiding place to another. Strictly on the basis of their prototype, however, Robot Defense Systems already has become the first company to develop a working, autonomous battlefield robot that can follow a preprogrammed path and plot new courses based on obstacles encountered.

Robot Description

The Prowler (Fig. 4-14) resembles a military armored personnel carrier, or a tank without a turrent. The prototype vehicle is 10 feet long, 5 feet wide, 4 feet high, weighs 800 pounds, and can carry a payload of 1000 pounds. It moves on six rubber-tired wheels driven by hydrostatic drive motors connected so that wheels on opposite sides can counter-rotate, allowing the Prowler to turn inside its own length. Under remote control (tested at 1200 feet), the robot runs at 10 miles per hour.

The production model robots, which also resemble tanks, have larger payload capabilities (sensors, television cameras, weapons, etc.) and carrying capacities of up to 1000 pounds; their 1200-pound weight includes a lightweight, composite-based armor skin. Mounted atop the 4800-pound heavy-duty basic platform, production model Prowlers stand 5 feet high. Two versions are available: an electric-drive "stealth" model powered by batteries, and one driven by a small diesel engine. The internal combustion version has a 250-mile operating range.

The Prowler can be operated manually via an umbilical cord, remotely via radio control, or under the direction of a semiautonomous on-board computer. The manual control option overrides radio control, which in turn overrides computer control. The remote control options feature full-color graphics display and joystick control of robot functions; communication is handled through an RS-232C interface and a high-speed 32-bit parallel link.

The on-board computer now uses a 32-bit, Motorola 68000-class microprocessor, which is to be upgraded later to the faster 68020 chip. The proprietary REACT software is based on a LISP-like language with a local dialect coded for real-time functions. Motor control and other lower-level functions use assembly or a structured language like Pascal or PL/I.

Fig. 4-14. The Prowler military and security robot standing sentry duty. Courtesy Robot Defense Systems, Inc.

66

Sensors (some mounted in a pod that can telescope up to 5 feet above the robot) include dual high-resolution television cameras with infrared night vision capabilities, ultrasonic sensors, a seismic monitor so that a stationary vehicle can monitor troop and vehicle movement, and armor impact sensors that can detect projectiles coming at the vehicle. Options include laser range finders and other distance measuring equipment, a gyrocompass, and an advanced land navigation system.

Defensive and offensive armament that the Prowler can operate include automatic weapons (such as the M-240, M-244, M-60, "mini-Gatling," and various small-caliber machine guns, as well as the Israeli-built Uzi), 20mm cannon, shotguns, anti-armor missiles, flame throwers, grenade launchers, tear gas foggers and grenades, Mace sprayers, sonic disrupters, and water cannon. The final determination of armament is made by the end user.

Industries and Applications

The two primary applications for which the Prowler was designed are surveillance and military operations. Military robots can be used to patrol sensitive locations such as air and missile bases, nuclear weapons stockpiles and other munitions storage sites. This security capability can be extended to include battle-zone reconnaissance and, as noted earlier, the robots developers also foresee a possible role for it as an unmanned tank destroyer. In the public/private sector, an appropriately equipped Prowler could be used for remote-control bomb disposal, or could be fitted out as a security robot to patrol banks, warehouses, industrial plants, prisons, mental hospitals, schools, high-security laboratories, private homes—almost any building or site with the minimal maneuvering room the robot requires.

In a military context, according to Dr. Mel Vuk of RDS, the Prowler has the advantages of responding predictably and doing exactly what it is programmed to do—even under heavy fire. It is physically small and lightweight, and is "survivable" because of its low silhouette, long-range mobility and maneuverability, and minimal infrared and radar signatures. It is a well-armored vehicle, protected against environmental factors, that also offers easy field maintenance due to its rugged design and on-board diagnostic systems. It is recallable, reusable, and (ultimately) expendable.

Because of the above factors, says Dr. Vuk, the Prowler can play a significant role in expanding the capability of obtaining combat information on:

1) The disposition, composition, and movement of enemy ground forces.
2) The location of enemy lines of communication, installations, and sources of electronic emissions.
3) Weather and terrain.
4) Conditions in surface battle areas.
5) Post-strike damage assessments.

This reconnaissance role minimizes personnel involvement, allowing humans to spend their time analyzing data instead of fighting. Used in conjunction with military air services, the Prowler robot can extend the normal range of electronic warfare systems so that field commanders can see and attack deep. The Prowler can provide near-real-time delivery of combat information and, as an autonomous robot vehicle, can be used immediately for tactical execution. It can enhance the electronic warfare battle environment by providing ground-based surveillance, jamming, and deception capabilities to the airborne components of a military strike force.

The Prowler also lends itself admirably to a wide range of sentry duty applications. It can be configured with the sensors necessary to detect intruders (including infrared imaging that allows it to see in total darkness), can interact with passive sensor systems, and can be monitored via closed-circuit television from a remote location such as a guard house.

Once a sensor signal is received, the robot will move in to investigate and assess the nature of an intrusion. If so programmed, it can respond to the intruder with nonlethal actions such as floodlighting the intrusion area and delivering voice-synthesis messages to retreat or hold still, and then escalate to disabling actions using sonic disruptors or chemical sprays such as Mace. Lethal response is also possible. Although the robot is capable of autonomous response, it may be programmed to allow human intervention before deadly force is employed. A guard house employee thus could make the final decision of what should be done at an intrusion, choosing to let the robot proceed as programmed or intervening by remote control.

RDS claims that no special skills are required of the operator, except human responsibility.

The industries for which the Prowler is targeted include the military and security fields. Although Robot Defense Systems has admitted that, unlike more utilitarian devices such as welding or painting robots, there is no guaranteed market for its Prowler, they have been active in the search for clients. As mentioned earlier, discussions have been held with United States Air Force and the government of Israel. According to *Robot Insider* newsletter, Robot Defense Systems also has signed a cooperative bidding agreement with Bechtel National, a unit of the San Francisco-based Bechtel Group experienced in international construction, to pursue security system contracts in an unnamed Middle Eastern country. The project, shrouded in secrecy, originally attracted 60 firms which put up $5000 each in bid bonds; RDS now will be bidding against 17 other companies for the contracts. Robot Defense Systems and Bechtel are the only team proposing robotics.

Summary

Name:	Prowler
Manufacturer:	Robot Defense Systems, Inc. 3860 Revere Street Denver, CO 80239 (303) 373-4984
Physical Description:	Small, 6-wheeled vehicle shaped like a military armored personnel carrier, weighing 800 to 1200 pounds and standing 4 to 5 feet tall. A variety of sensory hardware and armament is available.

Primary Use:	Patrol and sentry duty, hazardous travel, reconnaissance and tactical surveillance, transportation, decoy activities, mine detection and laying, search and rescue, and weapons deployment.
Significant Sales Point:	The Prowler can operate without radio control links in many types of hostile environments.

ROBBIE

In September, 1983, the Canadian firms of Cam Flow introduced Robbie, the cotton candy vending robot. Two Americans, Howard Danzig and Steve Shapiro of St. Louis, Missouri, discovered Robbie on a trip to Canada and first displayed him in the United States at a Business Opportunity show in Washington, DC, on December 3, 1983.

Robo-Vend, a St. Louis distributor, sells Robbie in the U.S. for $3995. To promote the Cam Flow robot, Robo-Vend placed it in amusement parks, shopping malls, supermarkets, and other places with substantial pedestrian traffic. For 50 cents Robbie would spin cotton candy on a stick—while playing music and blinking his red light "eyes."

The idea that "people can't ignore a robot" paid off. With his friendly personality, Robbie became the first vending machine robot that appealed to all ages. In one test location, 81 percent of all in-store traffic stopped to play with the robot; of those who stopped, 77 percent tried the robot. Out of 300 adults sampled, 93 percent said they would buy cotton candy from a robot again.

Robot Description

Robbie the robot is a coin-operated cotton candy vending machine; his primary function is to mix ingredients, cook cotton candy, and dispense the candy to customers (Fig. 4-15). Robbie has a head with a smiling clown face sitting atop a fiberglass body with a transparent acrylic door and an interior bowl, where cotton candy is spun while the customer watches. The body shape is circular, rather like a can standing on two small pegs. Two removable wheels can be attached to Robbie's "feet" for transport. Arms at the robot's sides are fixed to the body and do not move.

Robbie is 4 feet 10 inches tall, 1 foot 10 inches deep, 2 feet 3 1/2 inches wide, and weighs 77 pounds. The robot's electrical requirement is 700 watts, using 120- or 240-volt 50/60 Hz wall current. Its motors are heat-proof, and an electrometer shows the heat degree for temperature adjustment. Robbie is equipped with a sugar container that holds 1500 grams, releasing the sugar in 15-gram increments to the spinning bowl, which wipes clean with a damp cloth. A coin acceptor that rejects false coins is provided, and RoboVend offers a 90-day warranty on all electronic, electrical, and mechanical components.

Applications

Robbie, the robot, is suitable for theaters, supermarkets, arcades, shopping malls, grocery and convenience stores, restaurants, amusement parks, schools, hospitals, nursing homes, parks, airports, bus and railway stations, hotel lobbies, bowling alleys and billiard parlors, carnivals, stadiums—in short, almost any location where people gather or pass by. It also can be used for promotional purposes at trade shows, exhibits, fairs, and conventions.

Users and Operations

Children use the robot most frequently; if the children are accompanied by adults, the adults will often use it also. The user simply drops a coin into a slot on the robot. Amusing music plays and the red-light eyes immediately begin flashing as the candy is being made. In 70 seconds the cotton candy is done, the machine automatically turns off, and the customer removes his or her portion. Figure 4-16 is another shot of Robbie in action.

Existing locations report averaging $300 to $400 per week; in high traffic areas the robot has earned as much as $25 per hour. Each serving of cotton candy costs approx-

Fig. 4-15. Robbie, the Robot, is an entertainment robot used for promotional purposes. Built on the premise that "nobody can ignore a robot," Robbie makes and serves cotton candy while playing tunes and flashing his lights. Courtesy Robo-Vend, Inc.

Fig. 4-16. Two more satisfied customers of Robbie, the Robot. Courtesy Robo-Vend, Inc.

imately 3 to 4 cents to make and brings in 50 cents or more. Robbie the vending robot has become a success because he appeals to children—and to the child in all of us.

Summary

Name:	Robbie the Robot
Distributor:	Robo-Vend, Inc. 2138 Woodson Road St. Louis, MO 63114 (800) 325-4767
Physical Description:	Circular fiberglass body 4 feet 10 inches tall, 1 foot 10 inches deep, 2 feet 3 inches wide; weight 77 pounds, Integral cotton candy spinning and vending equipment.
Primary Use:	Coin-operated vending machine.
Significant Sales Point:	An entertainment robot that uses a space-age idea to attract people and sell a product.

THE ROBOT FACTORY PROMOTIONAL/ANIMATED ROBOTS

The Robot Factory of Cascade, Colorado, has been designing and manufacturing promotional and entertainment robots for 17 years. The manufacturer currently offers standard model robots, animated characters, and custom units. In 1966 The Robot Factory's first robot, an 8-foot-tall giant named Commander Robot, made its debut as an ice skater for the Ice Follies; by 1968, Commander Robot had been joined by eight more skating robots. In 1971 the manufacturer presented a non-skating robot on "The Merv Griffin Show." That robot has since been joined by more than 200 other robots in the United States, Canada, Mexico, Europe, Africa, Australia, and Japan.

All mobile robots manufactured by The Robot Factory are remotely controlled. The robots walk, roll around, talk, and flash lights. They are available with such functions as hat tipping, blinking eyes, and pincing hands. The robots are equipped with tape cassette decks for music, sound effects, or prerecorded messages. They are capable of inflating balloons, dispensing business cards or brochures, signing autographs, and making noise with bells, air horns, beepers, whistles, and sirens. Some models come equipped with video cameras, monitors, recorder/players, and games. Base prices for standard models range from $3750 to $15,000.

Computer-programmed furry animated characters joined the product line late in 1982. They perform musical shows that are preprogrammed on cassette; they also can be operated by a human to perform live, "real-time" shows. Standard and custom characters and shows are available. The company offers the characters, whose base prices begin at $6975, for portable and permanent installations.

In January 1984 The Robot Factory introduced a user-programmable experimenter's and educational robot with an on-board computer and sonar. Programs, including one for speech, are available for this robot in both do-it-yourself and ready-to-run form.

Robot Factory robots have programmed "jobs," i.e. they are singers, dancers, actors, comedians, radio and television personalities, sales robots, teachers, and therapists. They have appeared on such television programs as "Buck Rogers," "The Jackson Five Show," and "PM Magazine." They also have been used to make commercials promoting everything from video equipment through computers and shoe polish to yeast. These robots have endorsed political candidates. They have worked at trade shows for a wide range of companies. They entertain at amusement parks, convention centers, remote broadcasts, resorts, restaurants, fairs and expositions, and are used heavily in stores and shopping malls.

Robot Description

The *Robodroid* looks like a computer keyboard built into an armless, legless, three-cornered body with a jack-in-the-box head. Three openings in the head and one in the body look like speaker grilles. The robot is 32 inches high, 18 inches across, and weighs 35 pounds.

This robot uses guidance to roam on its own, speaking as it moves; its actions can be programmed with its on-board computer or with a joystick. Sensory input to its guidance system comes from three standard sonar transducers mounted on the rotating head, a head position indicator, and precise odometers on two of its wheels. Optional inputs include two remote joysticks, two paddles, and a light pen; future sensors slated to be added include an infrared detector, smoke and fire detector, light detector, radiation detector, carbon monoxide detector, burglar alarm, and digitized video I/O.

In addition to the two drive motors and the head motor, the Robodroid's control outputs can activate a sound/music synthesizer, a speech synthesizer with an unlimited vocabulary, and two real-time clocks. Future options will include an arm and a wagon.

This computer/robot (suitable for the hobbyist, educator, or experimenter) uses the well-known 6502 microprocessor and offers 64K RAM in addition to its 28K of read-only memory. The I/O system features a full upper- and lowercase keyboard with function keys, as well as eight ports that include two joystick ports, and audio/video port, cassette port, user port, microprocessor expansion port, and a board edge port. An RS-232C serial interface port is standard; and Centronics parallel port is an option.

A disk drive, cassette drive, printer, modem, and monitor round out the hardware complement. Available software includes the standard Robot Extended BASIC, a text-to-speech conversion program, and a joystick control program. Optional software includes a light pen control program, the CP/M operating system, and the languages FORTRAN, COBOL, Pilot, Logo, and Forth.

Another Robot Factory robot, the *Canbot*, looks like a soft-drink can with "Pepsi-Cola" printed across its silver-trimmed white plastic body, which is covered in magnetic-sign material. The robot features arms and hands and a domed head; it is 50 inches tall, 19 inches in diameter, 26 inches across the shoulders, and weighs 85 pounds. The Canbot moves forward, backward left and right, and spins. The arms move at the shoulders, the eyes blink, and it talks with the aid of a cassette deck. The robot is equipped with a business card dispenser; optional extras include a tipping hat and pincing hands. A traveling crate is included, as is a magnetic body cover/sign.

The *Milk Carton Robot* talks, moves forward and backward, turns left and right, and spins. Like the Canbot, its arms move at the shoulders, it comes equipped with a business card dispenser, and a cassette deck for music or messages is built into the body. The robot is 47 inches tall, 20 inches square, and 29.5 inches wide at the shoulders; the dimensions are scaled to the proportions of a standard half-gallon milk carton. The body is fiberglass and plastic printed in color, with the word "Milk" emblazoned across the body.

The *Videobot* talks, moves forward, waves its arms, and does everything else the other Robot Factory robots do. In addition, it contains a video recorder/player, monitor, and camera. Options include an air horn/balloon inflator, autographer, additional cassette decks, prerecorded cassettes, cigarette lighter, clock, brochure/card dispenser, voice-activated lights, pincing hands, sound effects, talkback, and voice modulation. This robot, made of stainless steel with a white plastic dome, stands 47 to 51 inches tall and weighs up to 125 pounds, depending upon how it is equipped; the body is 21 inches in diameter and 28 inches wide at the shoulders.

With its stainless steel body and plastic dome, the *Six T. Robot* resembles the Videobot, but is slightly smaller and has a hat lift. The radio remote control transmitter is concealed in a shoulder bag, and the robots "talk" by means of a wireless microphone that is invisible on the operator's body. The manufacturer claims that the Videobot and Six T. Robot are the most sophisticated entertainment/promotional robots in the world.

The *Hot Tot on a Tricycle* talks, walks, moves forward or backward, turns left or right, moves its arms and legs, and can ride a tricycle, three-wheeled motorcycle, bicycle with training wheels, and other vehicles. This robot, covered in cuddly fake fur, is suitable for a school or children's store, or even a private home. Although the dimensions vary with the type of riding vehicle, the body size of the Hot Tot is matched to that of the average nine- or ten-year-old.

The Hot Tot on a Tricycle sells for $4500, the Milk Carton Robot is priced at $5750, Canbot sells for $5825, and the Six T. Robot costs $7995. The Videobot, priced from $10,495 to $12,195 comes with a color VCR, monitor, and optional camera. The plain white dome and plug-in eye unit is priced at $265. Sealed batteries are used in all the robots to prevent acid spills, and each is delivered with a 6-month parts and workmanship warranty.

Industries and Applications

The main function of promotional and entertainment robots is to capture the attention of an audience in order to offer a repetitive advertisement or sales pitch—and offer it in a way that is significantly different than one's competition.

A robot in an elementary school, store, or pediatrician's office can keep children's attention while teaching them a lesson, bringing humor and fun into a task usually performed by a human. A flyer or business card given to a person by a robot might be remembered longer than one passed out by a human. The obvious psychology at work here is that "you can't ignore a robot."

Combining show business and electronics, The Robot Factory's furry animated and metallic robots are designed to entertain customers and promote businesses; the manufacturer also will design a robot to fit a special location or need. Used at commercial and entertainment locations such as stores and shopping malls, a robot—especially one tailored to help reinforce the user's message—can hook the audience's attention and capture an immediate customer crowd. In a very short time the robot can come to be identified as the company's mascot, product, or even the star salesperson!

Dow Chemical, Pepsico, Bell & Howell, IBM, Honeywell, A&M Records, Ice Follies, Holiday on Ice, American Airlines, Sony, Universal Foods, Teledyne, Haggar Slacks, Burger King, Philip Morris (Germany), are just some of the industries which have used these robots to advertise their products, not to mention scores (perhaps hundreds) of amusement parks, radio and television broadcasters, trade show producers, marketers, and advertising agencies.

Industrial use is almost unlimited, spanning almost every instance where advertising, promotion, entertainment, or educational messages must be put across. Although The Robot Factory's products are not "industrial robots" (in the sense that they do not weld, spray paint, or assemble parts), they could be used to deliver information such as safety lectures to workers. Since, in addition to long- and short-term rentals, the manufacturer offers the robots for sale, they also serve as the nucleus of an individual rent-a-robot business.

Users and Operations

Operating any of The Robot Factory's robots is simple. Short-term rentals include a human operator; a purchaser can buy preprogrammed software or create custom programs in a choice of computer languages. The "interactive" robots talk via a wireless microphone concealed on the operator's body, with the radio remote control equipment hidden in a shoulder bag.

The purely "entertainment" robots (the Animated Hot Tots, for example) have manual operation options, but do their shows primarily from cassettes programmed in advance. The Animated Hot Tots sing, dance, tell stories, and play instruments; one 60-minute tape will hold eight to ten shows. A user can have the robot perform standard shows, themed shows, or holiday shows; customized programs are available. The Animated Hot Tots, of adult-human size or larger, can have their fake-fur heads and costumes altered, or they can be made to wear human clothes—opening the way to costume them as cowboys, Santa Claus, or what ever else the user might require.

Summary

Name:

Animated Hot Tot
Videobot
Six T. Robot
Robodroid
Canbot
Milk Carton Robot

Manufacturer:

The Robot Factory
P.O. Box 112
Cascade, CO 80809
(303) 687-6208

Physical Description:

Varies widely. Average height is 50 inches, width 28 inches, weight 85 to 125 pounds. Robots can be configured to look like furry animals, milk cartons, and cola cans; also can be customized to resemble humans.

Primary Use:

Entertainment and promotion.

Significant Sales Point:

Eye-catching, effective promotion based on the principle that "you can't ignore a robot."

TRALLFA TR 3005

In 1964 Trallfa, a small Norwegian agricultural equipment company, started the concept of using robots to manipulate processing tools. In particular, the Trallfa painting robot was invented to sidestep the problem of recruiting Norwegian farm workers to do a particularly dirty industrial job. Ole Molaug, a talented engineer, produced an automatic painting machine, and by 1966 the first prototype robots were set to work painting wheelbarrows in the Trallfa plant in Byrne, Norway.

By 1969 Trallfa also had sold two painting robots to a Swedish factory to enamel bathtubs and showers; these robots are still working today. In 1976 the Trallfa robot, previously employed in England (as elsewhere) almost exclusively for spray painting, found a new use at the Ipswich factory of Ransome, Sims and Jefferies, another manufacturer of agricultural equipment. Ransome's, working with Trallfa engineers, applied the Norwegian robot to arc welding tasks with history-making results.

The Trallfa robot of 1964 had a wrist joint that could be operated in two different directions to move a spray gun within a three-dimensional pattern. There were a few initial problems making the robot explosion-proof and protecting it from the high tension of the electrostatic spray guns, but Molaug solved these problems through minor modifications to the manipulator and hydraulic pump. Later a sixth rotary motion for the spray gun was needed, so a new flexi-arm or wrist joint was constructed. The spray gun now moved in three directions—like the human wrist.

As robot control devices improved, Trallfa began designing its own control systems using resistor-transistor logic (RTL), the integrated circuits for which continued to drop in price and become smaller. Later Trallfa created a servo system that used an Intel 8088 microprocessor to store data on flexible disks and refeed it to the robot. A decade later, progress led to the design of a new control unit which uses a 16-bit processor and can control eight servos that provide extra mobility for the robot.

Equipped as a welding tool, the Trallfa robot today sells for about £ 35,000 for the robot and welding equipment, with a rotary mounting table costing an additional £6000. (At the early-1985 exchange rate, £1.00 equals approximately $1.10.) In the United States the Trallfa is made and distributed as an automatic spray-painting robot by the DeVilbiss company of Toledo, Ohio.

Robot Description

At first glance the "teach-by-doing" Trallfa robot resembles a dentist's drill and instrument tray, and consists of three parts: the manipulator (the working end), the hydraulic power unit, and the control center.

Rising from a platform-mounted base, the manipulator consists of a vertical arm connected to a horizontal arm at a movable joint. At the end of this horizontal arm is a wrist joint and a stub to which can be attached the paint spray gun or welding device. The arm and wrist actions allow six axes of motion, and wrist rotation is possible. The robot can be mounted on a slide base for spraying the interior of automobile bodies or stationary large objects, and two servos can be used to allow the robot to paint objects that are taller or wider than itself.

Two types of control center are available. A single cassette tape reader (SCT) contains a memory designed for use when identical parts are to be sprayed or welded, or when production consists of batch quantities of different parts. Each tape cassette holds one program for up to 85 seconds. The other type of control center is the Computer Robot Control (CRC) system, which combines a microcomputer with a dual floppy disk storage system and an LED output display.

The CRC module allows storage of 64 separate continuous-path or point-to-point motion programs, any one of which can be recalled in 0.5 seconds and can include automatic color changes. This rapid access to many different painting programs makes the system ideal for spraying parts loaded randomly on a conveyor; the manipulator motions can be sychronized with the conveyor via a pulse generator to compensate for speed differences. PROM memory can be added to the control module to allow linking of programs and to extend the robot's memory of parts it can identify automatically.

Industries and Applications

The Trallfa robot has been found exceptionally effective in industries engaging in tool manufacturing, electric welding, paint spraying and other coatings application, and automobile fabrication. Two noteworthy installations we will examine here are in the manufacture of agricultural equipment and the preparation of space vehicles by the National Aeronautics and Space Administration (NASA).

As a result of the publicity surrounding the Trallfa and it other Norwegian competitors, in 1976 the English agricultural equipment firm of Ransome, Sims and Jefferies began to look at the use of robots in their manufacturing operations. One potential application seemed suitable for the Trallfa: arc welding. Robots made by Trallfa and ASEA were tested; the Trallfa was chosen because it seemed better adapted to the particular task Ransome's had in mind—welding the part of a plow known as the "frog"— and because its "teach-by-doing" programming capability appealed to the welders who would be using it.

Trallfa engineers first set up the machine at Ransome's in 1976 and trained the personnel who would be handling the robot. Problems arose during the first 6 months of operation in the positioning of parts to be welded, as well as with the cassette-based robot control system then in use. Within the year Trallfa supplied a later version of the machine with a much larger, disk-based memory and control module. A rotary work table for the robot also was added to improve the motion efficiency and safety of the system. Since 1977, when they were placed in regular service, the Trallfa robots have allowed the company to improve welding consistency and realize great manpower savings.

Each Trallfa robot is used in batch production at a single work station, working as a team with a human operator who prepares the parts for welding and who can constantly check the robot's accuracy. Despite the possibility that this human pacing could limit the robot's productivity (as well as the fact that the company did have to spend the first year debugging the system), each robot now performs the work of at least two human welders—a figure which approaches three in practice because the robot needs no rest—while requiring only an operator and the services of a skilled welder for initial programming. On the basis of shift working and allowing 30 percent of the purchase price to cover debugging costs, Ransome's estimates a payback period of 2 1/2 to 3 years.

The National Aeronautics and Space Administration (NASA) uses Trallfa industrial robots in fabricating the space shuttle. The Marshall Space Flight Center applied several industrial robots to refurbish the shuttle after the initial flights. In 1980 the Trallfa was used to spray-coat the spacecraft with a protective coat of an epoxy-terminated urethane resin containing organic fibrous ingredients for strength, as well as a spray-on foam. The spraying was done once for ablative heat shielding and once for insulation purposes.

A spray also was developed for the reusable solid-fuel rocket boosters. This job called for a Trallfa T2000 robot, and electrically driven rotary table, and a Digital Equipment Corporation PDP-11/23 minicomputer. The spray path was pre-stored in the T2000 using the robot's continuous-path programming feature. A solid rocket booster spray system using Trallfa robots also was implemented in 1980 by United Space Boosters Incorporated at the Kennedy Space Center in Cape Canaveral, Florida. This system used the robots to spray larger items in several stages.

Users and Operations

With the Trallfa robots, a skilled welder or spray painter need be called in only for initial programming; once this has been accomplished, a relatively unskilled operator can be used to mount parts and monitor the quality of the robot's performance.

To program the robot, the operator attaches a teaching handle and function control box to the manipulator arm, loads a disk into the memory unit, and sets a console switch to programming position. A limit switch or photocell senses the presence of a part; when a signal indicating proper positioning is received, the programmer/operator begins to spray the part or perform the appropriate welds. The robot's *cycle time* (the time required to perform the complete sequence of operations) is longer than the human's, so that continuous robot operation is achieved. Where human preassembly time is needed, the user doesn't operate the robot at faster speeds.

In welding service at Ransome, Sims and Jefferies, the Trallfa robots have been used almost exclusively for welding agricultural equipment by operators who are familiar with assembling welding parts or who are themselves welders. Parts batches range from 100 to 500 units per run, with all parts of identical size in any one batch. A skilled welder programs the robot's end-of-arm welding motions at the start of each batch, and the moves are repeated immediately by the robot for the remainder of the batch, with a cycle time of approximately six minutes. Although they vary in size, the parts are similar for all models of plow; because of this, between-batch reprogramming presents few problems.

With respect to parts positioning, the system was originally set up such that welding was taking place at one station while parts were assembled by a human operator at another. When welding at the first station had been completed, the robot arm would swing across to the second station and the sequence would begin again.

The addition of the rotary parts-mounting table mentioned earlier streamlined the process and also improved worker safety by eliminating the potentially hazardous arm swings. Parts for welding now are placed by a human operator in special jigs on one side of the rotary table. The table rotates 180 degrees and the robot begins welding. At the same time the human operator assembles the next set of parts on the opposite side of the table and then presses a control button. If the robot has finished welding, the table rotates and the sequence begins again. The cycle times for mounting and welding are similar; the robot cycle is made slightly longer to allow for human functions while maintaining a high *arc-on time* (i.e., time spent actually welding) for the robot.

The mounting and programming methods are easy and provide great positioning reliability and repeatability. The quality of the actual welds is consistent, and the robot can produce work as good as the skills of the welder teaching the robot. Position drift of the robot arm does occur, however, at which point the run must be stopped and a human welder must be called in to reprogram the robot's motions. As noted earlier, the human operator of the robot need not be a skilled worker, but must be able to recognize faults such as misaligned welds. The user must then call a foreman to check the robot's setting or programming.

Interestingly enough, to achieve consistency Ransome's had to go back and improve the accuracy of earlier machining operations, because the robot can use only steel parts mounted in a jig according to position holes in the part. The accuracy of these holes is critical for consistency, and taking greater care in earlier machining operations did cause somewhat slower pre-welding parts production.

Summary

Name:	Trallfa TR 3005
Manufacturer:	DeVilbiss Division Champion Spark Plugs 300 Phillips Avenue Toledo, OH 43692
	Trallfa P.O. Box 113 N 4341 BRYNE Norway
Physical Description:	Horizontal arm (with wrist) 4 feet 5 inches long, supported by vertical arm 2 feet 7 inches, mounted on base. Resembles a dentist's drill. Total height 5 feet 8 1/2 inches. Six distinct motions.
Primary Use:	Spray painting and arc welding in manufacturing environment.
Significant Sales Point:	Manpower savings due to high productivity; consistency of painting or welding operations.

UNITED STATES NAVY UNDERWATER ROBOTICS

Funded by the U.S. Geological Survey, the Naval Ocean Systems Center (NOSC) in San Diego, California, has developed the unmanned, untethered submersible robot shown in Fig. 4-17 to provide research, location, inspection, and monitoring of underwater cables, pipelines, and structures. Applications of the vehicle include pollution prevention, enhanced safety in offshore pipeline and drilling operations, and search and recovery.

Weighing 4000 pounds, the underwater robot is 9 feet long, 20 inches wide, and 20 inches high. Easily launched from shore or shipboard, it is operational to a depth of 2000 feet and can run for approximately an hour with energy supplied from lead-acid storage batteries. The robot, called the "Free Swimmer," follows a set of predetermined program tracks such as parallel-path searches or figure-eight runs. A fiber-optic link for communication with the vehicle deploys as the Free Swimmer maneuvers underwater.

The vehicle may be equipped with motion picture or television cameras, along with the necessary underwater lights. Real-time video signals can be transmitted from the television camera through the fiber-optic communications link; an acoustic link has been

Fig. 4-17. With funding from the U.S. Geological Survey, the Naval Ocean Systems Center developed the Free Swimmer submersible robot to perform underwater search, location, and inspection tasks. Courtesy United States Navy.

Fig. 4-18. CURV III, the Naval Ocean Systems Center's cable-controlled underwater recovery vehicle, was developed to perform a variety of inspection, ocean engineering, search, salvage, and recovery tasks. Courtesy United States Navy.

Fig. 4-19. The Naval Ocean Systems Center's CURV III, shown here in its natural habitat, is capable of operating at depths up to 10,000 feet. Courtesy United States Navy.

demonstrated in separate experiments. A separate magnetic pipe sensor has been developed to allow the Free Swimmer to follow metallic pipes autonomously. The sensor is attached to the robot for deployment; its small size and weight do not impede the vehicle's maneuverability.

NOSC personnel are conducting ongoing research with the Free Swimmer system in the areas of fiber optics, advanced manipulator concepts, and supervisory-controlled electronic configurations. Remote operator systems and a voice-controlled manipulator with five degrees also are being investigated.

Only slightly more conventional than the Free Swimmer is NOSC's Cable-controlled Underwater Recovery Vehicle (CURV III), shown in Fig. 4-18 and Fig. 4-19. This robot also was developed for a variety of underwater inspection, ocean engineering, search, salvage, and recovery tasks. Equipped with an extensible manipulator arm, underwater lighting, and other tools, CURV III can operate at depths approaching 10,000 feet.

Job Opportunities

The fastest-rising job opportunities during this decade are going to be in computer-integrated manufacturing, machine maintenance systems, and making and maintaining robots. Machine vision is the backbone of computer-integrated manufacturing. It is likely to be the high-growth industry of the near future.

The Bureau of Labor Statistics lists 20 of the fastest-growing occupations between 1982 and 1995 as occupations in which people will be working with machines. Machine-related occupations make up 12 of the 20 "fast-track" job classifications. A recent Labor Department survey of 145 industries reveals the service industries to be among the 20 percent most capital-intensive. This nation's business is currently spending $25 billion a year on computer-integrated manufacturing. By 1992 that figure will leap to $100 billion, according to a report by Arthur D. Little, Inc.

Computer-aided manufacturing includes management systems that control up to 30 machines simultaneously. This field will grow to an $850 million market by 1990. Projections say that robots in U.S. factories will multiply twentyfold by the end of this decade, according to an April 8, 1984, *Los Angeles Times* article by James Flanagan, entitled "Robots Won't Take Jobs, Only Reshape Them." One robot does the work of up to six humans. More sophisticated robots will displace up to 3.8 million workers, according to a recent article in *Harvard Business Review* by Fred K. Foulkes, a university professor, and Jeffrey L. Hirsch, a labor lawyer.

Software can perform at a higher level than hardware. Due to this gap, many job opportunities exist with companies that manufacture robotics (peripheral/support) hardware and software. The machine vision/tactile sensor industry seeks to keep pace with the expanding software industry on which it depends. This constant striving to close the gap between hardware and software will cause job opportunities to proliferate. The focus today is on regearing the robot to meet the needs of existing but rapidly changing software.

Job opportunities exist for programmers, technical writers, salespersons, market researchers, and advertising executives to work for companies developing application software intended to run in the machine intelligence vision systems, which act as a robot's eyes. Opportunities exist in manufacturing process development laboratories, applied machine vision technology laboratories, and with users of programmable automation.

Technicians are needed to debug and compile applications of specific vision system control programs, to provide an application software test environment, and to duplicate production system environments so that a robot's function and timing may be modified. Some robot manufacturers have stopped marketing their line of robot systems and have turned to machine vision for more profits.

Programmers working with robotics machine vision systems run control programs written on machine vision development system. Jobs exist in the human factors field for industrial psychologists. These psychologists study human performance, training, safety, and human-robot integration.

Trainers are needed to instruct employees in new systems when older robots are upgraded with vision and tactile sensing systems. Job opportunities exist for vendors, tool servicers, planners and programmers, industrial engineers, maintenance technicians, managers, manufacturing quality control technicians, industrial relations personnel, public relations officers, employee relations counselors, toolers, computer repairers, robot operators, safety technicians, and legal specialists.

One of the biggest customers for robotics since 1970 has been General Motors. In the early 1970s, engineers from Unimation and General Motors worked together to produce the Puma assembly robot. More than a decade later, General Motors research is emphasizing vision systems at its Machine Perception Laboratory.

GM's robots are equipped with vision. At its Delco Electronics plant in Kokomi, Indiana, seeing robots inspect integrated circuit chips, testing at the rate of more than 3000 parts per hour. GM technologists are experimenting with voice recognition systems to enable a robot to respond to commands spoken by its operator.

Today GM has over 2000 robots in use in its plants. By 1990, that number is expected to grow to 15,000 to 30,000 robots. GM maintains a Robot Development Laboratory that recruits applications engineers from throughout the General Motors organization. In addition to engineers, the machine vision systems and robotics divisions of GM employs industrial sociologists, industrial psychologists, and other organizational development professionals who implement changes. General Motors and a Japanese robotics manufacturer, Fanuc, are currently engaged in a joint venture to manufacture industrial robots.

APPLICATIONS ENGINEERING

An ever-widening circle of robotics applications is creating a need for more application engineers and engineering technicians. Job opening exist in artificial intelligence and rehabilitative robotics. Artificial intelligence centers at Stanford University, Draper Laboratories, the National Bureau of Standards, and the University of Rhode Island are developing more applications for industry. Industrial opportunities for technicians, computer-aided designers/drafters, and programmers exist at industries such as SRI International in Menlo Park, California; General Motors; General Electric; and at companies such as Auto Place, Solid Photography, Object Recognition Systems, and Machine Intelligence.

The job opportunities in machine vision and applications engineering/technology exist for persons with liberal arts educations in marketing, sales, and advertising. Technical writing for manuals is one of the hottest areas in robotics today. Communications-oriented persons must write the scripts for the training films and compose clearly written instructional manuals. Opportunities in instructional technology, training, media presentation, and promotion are increasing in all areas of robotics.

The largest robot builders are the following: Unimation, Cincinnati Milacron, DeVilbiss, ASEA Inc., PRAB Robots, and Copperweld Robotics. Smaller companies are moving in fast, placing great emphasis upon machine vision/tactile sensor manufacturing. Job opportunities exist with companies that make value-added features for robots, rather than commodity-type products that compete by means of cutting prices.

The best opportunities are with firms that make computer controllers, software, and pre- and post-installation support products. Opportunities are good for field services such as on-site trainers and technicians. Job opportunities are increasing with distribution networks. These companies distribute robots produced by other manufacturers and make turn-key systems.

ELECTRONICS ASSEMBLY

Robots are increasingly being used to assemble computer components from disk drives to microchips. Intelledex Inc. stands out for its use of robots to assemble electronic parts. Unimation is now working on vision, tactile sensing, voice actuation, and mobility in its research laboratory in California. At Unimation focus has shifted away from hydraulics to more sophisticated electrical controls that involve fewer parts. This shift in emphasis is reflected in the type of employees Unimation might be looking for in the future. Technical people will be recruited increasingly from electronics engineering and mechanical engineering. Unimation will be looking for people who understand electronics from a control orientation.

ROBOTICS SOFTWARE

Software capability exceeds the mechanical resources at present. Technicians and engineers will be needed by such companies as Unimation, General Motors, and others to design new machines that can perform up to the capability of the software. The computer drives the machine faster than the hydraulic robot arm can react. Most jobs will go to the technician or engineer who can solve this type of problem.

The technology is already available in robotics today to solve these problems. Business depends on customer use. Unless a customer orders a specific item, most robotics builders cannot afford to invest time and money to develop a new system or discard the old hydraulic robot system for a newer electronic one that processes the tiniest of electronic parts.

Many U.S. robot manufacturers have Japanese partners. Unimation has Kawasaki. Westinghouse, involved with robots since 1971, brings its capital and clout to Unimation. Manufacturing has to cope with tremendous variety. This is a deterrent to productivity. Reducing variety inhibits business growth, but creating the capability to work with the variety creates flexibility and profit. Job opportunities are created when a company can respond rapidly to changes in the marketplace.

REHABILITATION ROBOTICS

The robot programmer is a person with a high school diploma and some technical training, and who knows how a robot arm is supposed to operate at a job site. Handicapped persons and those who work with them are opening new doors in robotics. The robots come in the form of vehicles, wheelchairs, artificial limbs, support braces, and other mechanical devices that enable the disabled to become mobile or to obtain objects from a robot by voice or breath command.

Job opportunities exist from time to time at the Rehabilitative Engineering Research and Development Center in the Palo Alto, California, Veteran's Administration Hospital. Physical and rehabilitative therapists, technicians, counselors, biomedical engineers and technicians, and occupational therapists are just a few of the workers involved in rehabilitative robotics.

At the NASA Jet Propulsion Laboratory in Pasadena, California, a robotic manipulator was mounted on an electrically powered wheelchair. A minicomputer-based voice recognition system was used to command and control the velocity of individual joints and to control the wheelchair. The Spartacus Project in France was the first to use a computer-augmented nuclear master-slave manipulator for rehabilitation technology. This was a robotic aid for the handicapped with optical proximity detectors on the terminal device.

Job opportunities exist with the Veteran's Administration, which supported the Stanford Robotic Aid, the first system to incorporate a human-scale industrial manipulator (Unimation PUMA 250). The goal of robotics technicians and biomedical engineers in rehabilitative robotics is to give handicapped persons robotic aids to help them in the activities of daily life.

Technicians work on telephones, food-preparing devices, personal hygiene mechanisms, vocational-task robotic aides, and recreational robotic aids. The robotic aid gives the disabled person the ability to manipulate objects so he or she can become independent. Designing interactive rehabilitative robotic aids is a rapidly expanding occupation. Technicians are needed to evaluate the robotic aid clinically and modify it.

Workers in rehabilitative robotics need to determine how the human user controls information to the robotic aid. Technicians and engineers must know how the robot processes new inputs and how the manipulative sequence can be made smoother for the disabled user. Occupational therapists must be able to evaluate how the robot's performance is communicated to the user. Physical therapists must relate the robot's power to human limbs and match robot movement to corresponding human movement, working closely with biomedical engineers and prosthetic technologists.

Opportunities exist for counselors to communicate with users of the robotic aid, helping the designers determine the need. Occupational therapy is becoming automated. Career opportunities can be found in the field of technology transfer; interested persons should contact the Veterans Administration-sponsored Rehabilitative Robotics Project in Palo Alto, California.

According to the newsletter *Robot Insider,* a quadriplegics' work station is to be contracted by the Veterans Administration Rehabilitation Research and Development Service in Washington, DC The work station combines a slow, simple robot and chin controls with a word processor and "feeding station" to help liberate persons with spinal cord injuries from confinement. The system was developed for the Veterans Administration by the Applied Physics Laboratory of Johns Hopkins University in Baltimore, Maryland. An advanced, flexible system is currently under development at the VA's Palo Alto, California, Research and Development Center.

AEROSPACE ROBOTICS

The defense industries, the military, and space-oriented industries such as NASA's Jet Propulsion Laboratory in California and the Johnson Space Center in Texas operate robots by remote control to obtain information from places hazardous or inaccessible to humans. Job opportunities exist in aircraft de-riveting and disassembly, parts assembling, welding, painting, machining, and tooling for technicians, drafters, engineers, programmers, and technical writers in the aerospace industry.

Jobs at military bases and with industries that have military contracts present opportunities for maintenance technicians and robot operators at all levels. Companies such as Hughes Aircraft, McDonnell-Douglass, and other manufacturers of aircraft utilize robots in production.

Robots fit into the factory environment in the aerospace industry in many functions. They are used for machine tool loading, spot welding, spray painting, casting, deburring, assembly, and inspection. The entire robotics field in a manufacturing environment is referred to as *computer-aided manufacturing.*

Students preparing for careers in aerospace robotics may choose to major in a two-year technician program specializing in either hydromechanical robotics systems or electromechanical robotics systems. Placement into robot-related jobs for graduates from this type of program has been very successful. One school which teaches these programs is Oakland Community College in Auburn Heights, Michigan. The Robotics Technology major is a two-year plus fourteen-week program leading to an Associate in Science degree. Courses available are robotics, hydraulics, pneumatics, CAD/CAM (development stage), integrated systems, microprocessors, maintenance/repair, and logistics. The program is open to any high school graduate. Persons seeking further information should write to the robotics contact person: Dr. Bill J. Rose or Edward Konopka, at Oakland Community College, Applied Technology Department, 2900 Featherstone Road, Auburn Heights, Michigan 48016, or phone (313) 852-1000.

COMPUTER-AIDED DESIGN AND MANUFACTURING

Industrial design and drafting has evolved into computer-aided design (CAD). The act of using computer graphics to create illustrations, designs, blueprints, and other printed and illustrated material has opened up one of the fastest rising industries in automated manufacturing.

Robots are becoming an important element of computer-aided manufacturing (CAM). These machines form a computerized network of computer-aided design systems. Robots connect each automated manufacturing process to a centralized computer, which develops

SIDNEY B. COULTER LIBRARY
Onondaga Community College
Syracuse, New York 13215

its database from a network of computer-aided design systems. Job opportunities exist where there is a centralized system of control which monitors and communicates with a distributed network of automated manufacturing processes known as *computer-integrated manufacturing* (CIM).

Jobs exist for parts designers, parts classifiers, and parts storers and cataloguers. Technicians in computer-integrated manufacturing study engineering technology in two-year community college and/or technical school programs to prepare for the many jobs at the technician level in the fields of CAD, CAM, and CIM. A student can pursue a four-year degree in industrial, electrical, or mechanical engineering, robotics, or electronic technology, or take a one- or two-year training program in automation technology, robotics, mechanical engineering technology, or industrial engineering technology. Courses based on the trend toward computer integrated manufacturing are offered at schools such as Wentworth Institute. CAD/CAM courses also are offered by Integrated Computer Systems, a private firm that teaches CAD/CAM technology overview courses throughout the nation for fees under $1000.

WELDING AND PAINTING ROBOTICS

DeVilbiss is one of the leaders in robotics manufacturing. Although the company probably is known best for the Champion spark plug, the firm turned the invention in 1888 of an atomizer by a Toledo physician, Dr. Allen DeVilbiss, into a multi-million dollar business. DeVilbiss equipment sprays paint on products, chocolate on crackers, stain on boots, vitamins on food, color on eyeglasses, and heat-resistant coating on the space shuttle *Columbia*. In 1982 DeVilbiss accounted for an 85 percent share of the spray-finishing robotics industry.

Job opportunities exist in welding and spray painting because software has run ahead of machine development. Salespersons in large numbers are needed in the robotics industry, to show customers why they need a robot system and how the expenditure will save them money.

Opportunities in welding are not always as welders. Robots do the welding and spray painting, but humans are needed to fill jobs in which they can convince a customer that the system can do the job. Technicians are needed with the expertise to install the system; programmers are needed with the expertise to integrate the system into the facility. Currently there is a glut of hardware and software designers; the need is for applications experts in welding and paint spraying robotics. Finding a technician with experience is very difficult at present. Firms such as DeVilbiss are looking for a person with electronics experience who can be taught the welding and spraying business, or someone with welding and spraying experience who wants to learn robotics. DeVilbiss draws about half of their service technicians from persons leaving the military forces. The company offers tuition aid amounting to 70 percent of expenses.

A recent article in *Occupational Outlook Quarterly*, a publication of the Bureau of Labor Statistics, predicted that maintenance workers will take the largest number of the new robotics jobs. There will be a great demand for programmers to develop robotics software for the CIM industry. For technician trainees in welding and painting robotics technology, the job situation is brightening. A robotics training program for unemployed and displaced workers graduated its first class in August, 1983, at the Control Data Business and Technology Institute in Toledo, Ohio. The Private Industry Council and

The Employment and Training Commission sponsored a robotics technician program at Career Works, Inc., an employee-owned career development firm. The 40 weeks of training included a week at Unimation, Inc., in Farmington Hills, Michigan.

In Dearborn, Michigan, the Society of Manufacturing Engineers started a $300,000 Challenge Campaign which doubles the value of corporate and individual contributions to its Manufacturing Engineering Education Foundation. The Society matches dollar for dollar up to $300,000 in all-cash donations to the Foundation. Almost everywhere in the nation, funds for training robotics technicians and engineers are being poured into programs by private industry.

Fully automated factories need people to work in computer-integrated manufacturing. Technicians are needed to monitor machining and metal fabrication. Jobs exist for automation inspectors, even though there are robots that inspect products because sensor technology is limited. Jobs for people in materials handling still exist because today's robots can handle only parts that are similar in shape and size or that can be transported on a standard pallet.

Job opportunities will be integrated into a CIM system. People—like software and equipment—need to be integrated, so that jobs in the human factors in technology area are going to increase significantly. Skilled welders and paint sprayers are needed to teach the robot to imitate their steps. Technicians are needed to diagnose and repair. They are the "machine doctors." Engineers, programmers, and paraprofessionals are needed to design and manage a CIM facility. Writers are needed to compile instructional manuals and training films. Trainers are needed to teach robotics technology.

Computer-integrated manufacturing uses hundreds of skills to integrate all the disciplines of technology, psychology, and communication. According to projections from the Society of Manufacturing Engineers, the robot manufacturing industry will employ between 60,000 and 100,000 workers by 1990. In the following section we'll take a look at the education and training available to prepare for some of these job opportunities in robotics, computer-aided design, computer-aided manufacturing, and computer-integrated manufacturing.

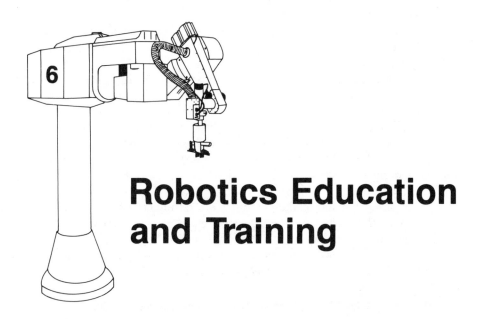

Robotics Education and Training

Training for robotics careers is conducted by robot manufacturers, users, educational institutions, technical societies, and film companies that specialize in robotics education. Robot manufacturers generally are responsible for two types of training on their robots: they provide training for user personnel and for their own employees.

Training for users is aimed at an overall understanding of operation, maintenance, programming and control, according to the Society of Manufacturing Engineers. Classes are usually aimed at teaching functions of operation and maintenance. This instruction consists of classroom discussion, visual aids, documentation, and hands-on experience. Training for employees of robot manufacturers consists of on-the-job work experience supplemented with the same instructions given to user personnel.

The instructors for robotics training programs are usually employees who have gained broad knowledge about their products through work experience with robots. Their formal education usually has resulted in at least an associate degree or more, plus in-depth involvement with the robots and equipment.

Training may be done in the user's plant by one of the user's employees, based on how many robot installations a user has or anticipates. Most plants with major robot installations maintain a training program. Instructors may update their knowledge by attending seminars and classes. Most instructors also have similar educational and experiential backgrounds to manufacturer's instructors. Multi-plant operations usually send employees to a central training center for instruction, or the instructor may teach at the plant location. Employee turnover is generally moderate, so that instructors have to repeat classes periodically.

Teaching robotics in higher education ranges from introductory overviews at community colleges to doctoral programs at universities and research centers. The focus on robots at all levels is related to the subject being taught. The basic principles of the subject in each course usually are applied to the robot as a working tool.

Many robotics training programs and schools are using individuals from industry to teach classes. One of the deepest needs in robotics is retraining workers seeking to become involved in robotics technology. Often a teacher will go out to a plant to conduct a class on robotics for employees.

The number of robot installations is increasing; growth will create more job opportunities. A wide variety of jobs exist in business, education, and industry for trained robotics workers. It is possible to train for a career in robotics without formal education in engineering. Technology training forms a base for various job categories. In addition, subject matter and robotics courses are available to the public through continuing skills programs offered by professional societies, such as the Society of Manufacturing Engineers workshops and classes.

Potential employers, robot users, robot manufacturers, and robot vision/tactile sensor systems designers offer on-the-job training or retraining. The coordinated effort of schools and industry is meeting the challenge to provide experienced robotics personnel. The following section provides a list and detailed description of some of the robotics education and training available. Appendix A provides a comprehensive listing of two-year, four-year, and graduate-level robotics training.

SOCIETY OF MANUFACTURING ENGINEERS

Robotics International of the Society of Manufacturing Engineers (SME) offers a schedule of clinics, workshops, courses, and other educational programs that offer certification or recertification credits to manufacturing technologists and engineers under SME's Manufacturing Engineering Certification Program. The Society is a professional engineering society dedicated to advancing manufacturing technology through the continuing education of manufacturing technicians, managers, engineers, and robotics trainees. The SME also maintains an extensive communications department for instructional technology, where it publishes books on robotics and disseminates information.

The Society was founded in 1932 as the American Society of Tool Engineers. From 1960 to 1969 it was known as the American Society of Tool and Manufacturing Engineers; in January of 1970 it became the Society of Manufacturing Engineers and added Robotics International, the robotics technology and engineering division of the organization.

The changes in name reflect the revolution in the profession of manufacturing engineering/technology, from tool and die making to robotics and machine vision design. Courses that reflect increasing sophistication are offered to train persons who wish to enter the robotics field or who are already employed in robotics and/or manufacturing.

The Society has 70,000 members in sixty countries, most of whom are affiliated with an SME chapter. There are currently over 300 local chapters, plus Senior Chapters. More than 9000 students are active in over 140 SME student chapters throughout the nation. SME is a member of the World Federation of Engineering Organizations.

To encourage careers in robotics and manufacturing technology, SME's Education Department provides a diversity of courses and programs. Resources include career guidance workshops and publications, a curriculum directory with over 200 schools listed, newsletters and education reports for educators, annual educational forums, and guidance to Senior Chapters in continuing education activities.

Since 1954 SME has supported with capital equipment grants a university-level research and educational grant program to assist the development of educational curricula in manufacturing. The Society has established the Manufacturing Education Foundation, which supplies grants to schools to upgrade equipment and to provide scholarships.

The Society, through its Manufacturing Engineering Certification Institute (MECI), certifies manufacturing technologists and engineers. MECI certification is based upon the individual's technical expertise, knowledge acquired through years of on-the-job experience, education, and examinations. Certification programs are available for non-degreed, degreed, and registered professional engineers.

A typical SME course in "Applying Robots in Material Handling" runs for three days and includes the following six areas:

A) Introduction to robots for material handling.
   1) Definitions
   2) Application requirements
   3) Robot classifications
B) Material-handling robots, components, and tooling
   1) Types of power systems
   2) Application requirements
   3) Robot tooling for material handling
C) Material handling systems
D) Integrating robots for material handling applications
E) Robotic/material handling problem discussion
F) Robot laboratory session
   1) Robot familiarization, operation, programming, interfacing, tooling, and safety
   2) Hands-on programming development experience

Continuing education is emphasized. By attending technical programs, those certified add to their knowledge and simultaneously earn required credits toward recertification. Since the certification program began in 1972, more than 19,000 manufacturing personnel and trainees have applied for certification in a variety of specialties. Over 11,000 persons currently are certified. SME works with the Accreditation Board for Engineering and Technology, Inc., to accredit manufacturing engineering technology curricula.

Courses and workshops are usually presented through combined conference and exposition. SME sponsors 30 to 40 such convention workshops annually. SME's annual meeting held each spring—The International Tool and Manufacturing Engineering Conference and Exposition—attracts hundreds of exhibitors. More than 100 technical presentations are made describing the latest applications in manufacturing, metalworking, and robotics technology.

Year-round educational activities include more than 200 three-day programs annually (clinics, seminars, and workshops). These short courses, devoted to a single subject, include technical paper presentations, problem-solving sessions, and plant tours. For individual and group instruction, the Society offers academic course materials on various aspects of manufacturing and robotics. These include programmed learning courses, home study courses, correspondence courses, videotape training programs, textbooks, and handbooks.

Typical courses offered include hands-on experience with robots. For example, in May of 1984 SME offered a workshop on applying robots in material handling and arc welding with industrial robots. The three-day session offered a hands-on workshop to train the participants to work with robots and related peripheral tooling and equipment. The students learned to evaluate and implement robots for material handling applications. Subjects covered were robot tooling, sensors, and material handling systems. Laboratory experience gave the participants a chance to work with actual robot operation. Robotic material handling work stations illustrated the various capabilities associated with programming and interfacing with the robot.

The robotic arc welding workshop emphasized techniques of increasing arc-on time (i.e., minimizing the time between welds) and weld quality. Students were shown improved software routines for arc welding and seam tracking systems. Laboratory experiences enabled the student to gain experience in programming and operating robots. The workshops are aimed at technicians, supervisors, and persons interested in learning to operate a robot.

Offered at:

Robotics International of SME
One SME Drive, P.O. Box 930
Dearborn, MI 48121
(313) 271-1500, Ext. 394

Time of Course:

Usually three days.

Tuition:

Course tuition averages $550 for members of SME and $610 for nonmembers. Tuition discounts are available to full-time students and faculty. A $50 deposit is required. Participants are asked to contact the school directly for tuition fee information on individual courses.

Robots:

General Motors Institute and other corporations cooperate with SME in providing robots. Also, SME provides grants to schools to purchase robots and equipment.

Certificate:

Certification is awarded upon application and the passing of examinations in specified fields. Certificates are awarded in 19 areas of specialization, and certification study guides are available. These guides prepare the student for the certification examination. Interested persons should write for the Certification Application and Information Brochure.

Credit:

University credit for workshops is awarded upon completion of necessary registration procedures and successful completion of academic course work requirements. For more information concerning university credit, you may write to the Society of Manufacturing Engineers at the address noted earlier, or call (313) 271-1500, extension 381.

Additional Services:

A block of rooms is usually held at a hotel for workshop attendees. SME also offers a list of optional hotels. Workshops are held in various cities nationwide to accommodate students in different locations. Tours of local manufacturing plants are often arranged.

UNIVERSITY OF ARKANSAS CENTER FOR ROBOTICS AND AUTOMATION

Training in robotics is given by accredited departments in the University of Arkansas College of Engineering, Fayetteville. Robotic studies majors are available in three disciplines: Electrical Engineering, Industrial Engineering and Mechanical Engineering. The titles of the courses are listed by departments. Specifically, the Center of Robotics and Automation opened in 1982. The bachelor's degree is offered in the three disciplines. Graduate degrees may be earned by additional concentration in robotics and automation, with an appropriate thesis, in one of the departments.

In addition to the formal classes which are offered by the College of Engineering at the University of Arkansas, seminars are given by the University in Fayetteville and in Little Rock, the largest urban center in Arkansas. Continuing Education Units are given for attendance in these seminars.

The following prospectus information has been reprinted with the permission of the University of Arkansas.

The Center for Robotics and Automation has been established by the University of Arkansas at its Engineering Experiment Station. The Center was created and designed for the express purpose of assisting industry—particularly small- to medium-sized facilities—to develop and implement the use of robotics and automated systems without the serious financial risk that so often accompanies the utilization of emerging technologies. The Center accomplishes this by providing development space and assistance, engineering expertise, student and employee training, information dissemination, and exhibition space for vendors. In addition, it provides a method for supplying the College of Engineering with an ongoing, state-of-the-art laboratory for training industrial representatives and both graduate and undergraduate engineering students.

Center Functions

The Center performs five unique functions, each of which is discussed below.

Industrial Research and Development. For member firms, the Center for Robotics and Automation provides a unique service: member firms may reserve space in the Center to pilot test and debug proposed installations. They may also test their own ideas, equipment, prototypes, or systems. In addition, development space is provided so that industry can test ideas prior to the purchase of expensive systems. Engineering faculty members and graduate students are available to assist in any or all of these research activities. This service can greatly reduce financial risk for smaller industries who might otherwise be unable to install a robot or automate any part of their manufacturing systems.

Vendor Services. The Center is equipped to test and evaluate member-supplied equipment; additional testing will be performed by mutual agreement. In order for the Center to have the best equipment available (both to test and with which to test other equipment, it either purchases, seeks donations, or borrows on a short-term basis. While member firms are providing funds to buy some equipment, there is a continuing need for the Center to seek equipment. This equipment is then integrated into research, seminars, and other educational efforts.

Firms placing equipment in the Center can be assured that their equipment will be the center of research efforts, in the hands of hundreds of practicing engineers, and in the hands of many hundreds of both undergraduate and graduate engineering students.

Joint Industry-University Research and Development. Firms interested in efforts within the Center which may involve more than the simple use of the laboratory or facilities are invited to propose joint efforts. In this manner member firms may work with the Engineering Experiment Station personnel (faculty members from the College of Engineering) in joint research and development projects. For example, a member firm may desire to test its robot at varying temperature and humidity conditions. This is accomplished easily through the use of the new environmental chambers and a properly designed experiment. New control languages can also be developed and tested and new applications simulated at the Center.

Such joint efforts serve industry and increase the exposure of the engineering faculty to new developments in the field. Member firms have available all of the expertise of the Center for Robotics and Automation on an as-needed basis.

Continuing Education. In an effort to encourage the use of robotics and automation, there is a need to expose practicing engineers in industry to the capabilities of current equipment and concepts. To meet this need, the Center continually conducts seminars and short courses. The Center has offered four seminars through 1983 and will continue to offer seminars, subject to demand, such as the following:

- Robotics and Manufacturing Automation (2 days)
- Economic Justification of Robots (2 days)
- Programmable Controllers (2 days)
- Introduction to Numerical Control (3 days)
- Industrial Robots (3 days)
- Automated Warehousing Systems (3 days)
- Automatic Inspection Stations (2 days)

■ Mechanical Manufacturing Systems (2 days)

Each seminar/short course (taught by both University and industry experts) is offered with a hands-on laboratory where appropriate. In fact, the development and existence of this laboratory and its exposure to practicing engineers is crucial to the success of the Center. Visitation to the Center will be possible at any time by appointment.

University Education. In response to the current crisis in engineering education, this Center has become a tremendous asset to the graduate and undergraduate engineering students at the University of Arkansas. Engineering educators must constantly develop state-of-the-art laboratories for those engineers interested in manufacturing. This has proven to be so difficult for universities that most have not succeeded.

By equipping and maintaining the Robotics and Automation Laboratory, the Center allows the University of Arkansas to educate every graduating engineer in the state-of-the-art manufacturing technologies. Existing courses which currently depend on the Center are:

■ Manufacturing Systems Design (Junior level)
■ Automated Production (Senior level)
■ Computer-Aided Manufacturing (Senior level)
■ Robotics Applications (Senior level)
■ Graduate courses in Automation

Center Membership

A membership fee of $7500 annually guarantees the following rights and privileges:

■ Use of the Center's facilities and staff for five working days or seven consecutive days.
■ Free testing and evaluation of member-supplied equipment (one per year) with additional testing according to industry specifications by mutual University/industry agreement. All tests will be documented by a confidential engineering report.
■ Two attendees per seminar or short course.
■ Free copies of Center publications as appropriate.
■ Company name on all Center publications if desired.
■ Other benefits as developed.

The annual fee of $7500 will cover the following items:

Equipment purchase	$2500.00
Equipment maintenance	750.00
Center personnel	1800.00
EES facilities	1250.00
Seminar costs	1200.00
	$7500.00

Center Facilities and Equipment

The Center has over 10,000 square feet of available floor space, three environmental chambers capable of extreme temperature ranges, and several large classrooms. In addition, the Center (currently) has available 11 robots ranging from the small experimental class to the very large heavy-duty class, which cover all major types: electric, pneumatic, and hydraulic. In addition, there are programmable controllers, over fifteen microcomputers from various manufacturers (Apple, TI, TRS-80, DEC, etc.), interface modules of wide ranges and descriptions, an Aerotech/Bridgeport numerical control mill, application models or simulators, and a 30 × 10 feet powered material conveying loop. Additional equipment is being acquired and member firms can be expected to supply even more.

Center Management

The Center is managed by a Directing Professor from the College of Engineering, University of Arkansas, Fayetteville. This Director manages the day-to-day operation of the facilities and personnel. The Director also provides the interface between the Center and all interested parties.

The Director is provided with policies and guiding philosophies from a Board of Directors comprised of the Dean of the College of Engineering, three chosen industrial representatives, the Department Heads of Electrical, Industrial and Mechanical Engineering, and the Director. The Board meets at least once per year to review the Center's operation and provide guidance.

Industrial Engineering Courses

Manufacturing System Design. This course covers the selection and design of productive systems, and of industrial processes and process sequences to manufacture products.

Automated Production. Covers mechanization and automation concepts, analysis of flow lines, logic control systems, introduction to numerical control and computer assisted manufacturing, and laboratory projects.

Computer Assisted Manufacturing. Course material includes off-line assist and on-line computer control of manufacturing processes; interfacing the computer with the process; programmable controllers; computer numerical control; laboratory projects.

Materials Handling. Covers equipment, systems, problems, and analysis of industrial material handling, with emphasis on manufacturing. Also included are vehicles, containers and racks, conveyors, overhead systems, and miscellaneous equipment, as well as criteria for selection and decision models.

Application of Robotics. Covers the background, structure, drive systems, and effectors, and the general applications of robots; students develop a major applications project in the laboratory.

Automated Systems. This is a graduate-level introduction to principles of automation, mechanization of factory processes numerical control of machine tools, open and closed loop control, logic circuit control, analog computer simulation techniques, and computer control systems.

Mechanical Engineering Courses

Introduction to Machine Control. This course provides an introduction to feedback control concepts with emphasis on mechanical and fluid systems. Basic systems dynamics, Laplace transforms, matrix algebra, systems diagrams, transfer functions, stability, state space techniques, computer simulation, and case studies of typical control of industrial machines, nuclear reactors, and mobile equipment are covered.

Machine Systems. A unified approach to a mathematical modeling and analysis of mechanical, fluid, thermal, and electric systems is provided.

Sensors, Transducers, Interfaces, and Microprocessors. This course features an introduction to the various sensors and transducers and their interfacing with microprocessors for data transfer and control, and includes classroom and laboratory exercises in which transducers and other electromechanical devices are interfaced with microprocessors.

Two additional courses, "Robotics and Manipulative Controls" and "Advanced Sensors and Microprocessors" also are offered.

Electrical Engineering Courses

Control Systems. This course includes mathematical models of control systems; performance criteria and stability; Zeigler-Nichols, root-locus, and frequency-response design techniques; and special topics.

Control Systems Laboratory. Course material includes experimental study of various control systems and components; control system simulation; measurement of system parameters, process-control applications, and electromechanical systems; and selected applications.

Offered at:	Center for Robotics and Automation University of Arkansas College of Engineering 309 Engineering Bldg. Fayetteville, AR 72701 (501) 575-3156
Time of Course:	The seminars may be two or three full days. The college credit courses are three semester hour courses. Bachelors degrees normally take 4 to 5 years to complete.
Tuition:	The fees are $390 per semester for in-state students and $600 per semester for out-of-state students. The seminar charges average $125 per person.
Robots:	There are eleven robots in the Center which include electrical, pneumatic, and hydraulic drives. The robots are as follows:

- Schrader-Bellows MotionMate Two Position
- Pneumatic Robot
- Cincinnati Millicron HT3-586 Heavy Duty
- Hydraulic Robot
- Two Minimover-5 Microbot Robots
- Three Microbot Teach Mover pendant controlled robots
- Rhino XR-1 Laboratory Research Robot
- TI 510 Programmable Controllers
- Bridgeport CNC Vertical Milling Machine
- Heathkit Hero-1 Ambulatory Anthropomorphic Robot
- Unimate Puma 560, 6-axis Servo Controlled Robot
- VSI, Automation, Charley 4, Scara-type Assembly Robot

Certificate:

Bachelors, Masters, and Doctor of Philosophy degrees are granted by the College of Engineering. Continuing education units are given for seminars.

Additional Services:

The University has an active placement office. Academic course credits are accepted at any engineering school in the nation.

DES MOINES AREA COMMUNITY COLLEGE

Des Moines Area Community College is a publicly supported two-year institution created in 1966 to prepare or retrain students for employment and advancement. The High Tech-Robotics and Process Control program prepares a participant for a career as a technician in industrial manufacturing/robotics. At the end of the program the student should be able to diagnose and repair robots and industrial equipment, ranging from the basic motor control devices used in hard automation to the more sophisticated industrial robots that utilize microprocessors for programming and servo control.

The course of study includes the fundamental technologies such as pneumatics and hydraulics, as well as system applications such as process control and robotics. Upon program completion, a student may seek employment with manufacturers maintaining plant equipment or with companies that produce process control or robotic devices.

In order to enter the program, the student must have the equivalent of two semesters of high school algebra. The prospective participant must attend a workshop and obtain a satisfactory score on an algebra aptitude test. The school recommends but doesn't require two full years of high school algebra. To qualify for the Associate in Applied Science (A.A.S.) degree, students must complete the courses with a grade point average of 2.0. The student attends for six terms.

The program prepares the participant for a career as a technician in industrial manufacturing. The student learns to diagnose and repair industrial equipment ranging from motor control machines to microprocessor-controlled industrial robots. The curriculum includes two semesters of hydraulics and pneumatics, and a course in motor control/process control instrumentation with laboratory classes.

Students take such courses as industrial electronics, robotics, robotics laboratory, mechanisms, mechanisms laboratory, digital circuits, data communication, and laser applications. Two semesters are spent on digital circuits, as well as several semesters of circuit analysis courses and laboratories. Participants also study a year of technical math, computer programming, technical report writing, and fabrication techniques.

Offered At:	Des Moines Area Community Collect 2006 South Ankeny Blvd. Ankeny, IA 50021 (800) 362-2127
Time of Course:	Six terms
Tuition:	Total tuition costs for the A.A.S. degree are approximately $2,823.40. Tuition is $27 per credit with a maximum of $405 per term.
Robots:	Student should call the toll-free school number for the names of the robot models.
Certificate:	The Associate in Applied Science degree is offered after six semesters of study.
Additional Services:	The two-year college provides counseling and placement services to students. Off-campus adult and continuing education programs are also offered. Adults are assisted in completing their high-school education.

INDIANA VOCATIONAL TECHNICAL COLLEGE

Industrial Maintenance is the robotics program offered since September, 1982, by Indiana Vocational Technical College. The courses given by this school cover what is expected in the field of maintenance in industry, according to the school.

In the course of two years, the curriculum covers everything that makes up a robot, plus areas that would support robotics. Examples of this are welding, plumbing, mechanical metrology, and mechanics. Classes offered on campus in the field of robotics are Rotating Machines, Hydraulics and Pneumatics, Programmable Controllers, Basic AC/DC, Basic Electronics, Digital Principles and Applications, Introduction to Robotics, and Automation Production and Systems.

The course of instruction is 105 credit hours per quarter, consisting of seven 11-week quarters. Robotics classes have one or two credit hours of hands-on training plus field trips in the appropriate areas of instruction. The school uses two Microbots and a Hero-1 robot for robotics training. They also use Apple Computers in conjunction with the Microbot. Students receive an Associate degree in Industrial Maintenance at the end of the seventh quarter. Participants may enter or exit the program at any quarter, or they may take individual classes without pursuing the degree. The school accepts credits from other schools, and there is a placement service at the college.

Mechanical Metrology presents instruction and laboratory experiments in the use of mechanical test and measurement employed in quality control. Industrial Safety covers the responsibilities of management and supervision toward attaining an accident-free organization. Topics covered include starting and stopping machines and accident prevention measures.

Electrical Wiring Fundamentals include the use of electrical measuring instruments, circuits, and basic electricity. Plumbing Fundamentals include basic home plumbing. Other courses include introductions to active devices such as vacuum tube and transistor devices. Rotating Machines introduces common industrial machines, with emphasis on power distribution. The student learns the national and local electric codes, electromechanical controls, and digital principles through use of Boolean algebraic expression.

Troubleshooting Skills is designed to introduce work procedures needed to conduct planned, breakdown, and unschedules maintenance on robotic equipment. Emphasis is placed on the use of logic in troubleshooting. Other courses cover advanced electromechanical controls, hydraulic and pneumatic systems and repair, industrial microprocessor fundamentals, and industrial digital principles and applications. Students take an industrial digital laboratory, basic drafting, industrial processes and systems, and introductory welding. Other courses include mechanics and technical mathematics.

Many electives are offered in such courses as plant layout, blueprint reading, electronics, semiconductors, microprocessors, industrial controls, machine diagnosis and repair (electrical), machining, technical reporting, and mathematics.

Offered At:	Indiana Vocational Technical College 3501 First Avenue Evansville, IN 47710 (812) 426-2865
Time of Course:	105 credit hours per quarter for seven quarters, each eleven weeks long.
Tuition:	Total fees excluding books and supplies are $2,053 for 91-105 credit hours re-

	quired for the Associate Degree in Industrial Maintenance. The cost per credit for Indiana residents is $20.95; out-of-state tuition is $39.20 per credit hour.
Robots:	Two Microbots and a Hero robot are used for training. Apple Computers are used with the Microbot.
Certificate:	An Associate degree is awarded after 91 hours of study. A Technical Certificate is awarded after 46 credit hours. An Occupational Certificate is awarded after 22 credit hours of study.
Additional Services:	The school assists students with available resources for financial aid and veterans educational benefits. The Placement Office provides assistance in conducting a job search. Recent placement statistics indicate that nearly 90% of the graduates are employed, with the majority having jobs in specific fields for which they were trained. A list of housing opportunities is available through the Office of Student Services. Counseling is available.

INTELLEDEX ROBOTICS CUSTOMER TRAINING

Intelledex, Inc., of Corvallis, Oregon, is a relatively new (1982) robotics manufacturer that offers factory, regional, and on-site customer training programs. Classes are conducted in a well-equipped classroom featuring a dedicated training robot for demonstrations and hands-on training. Factory tours are included in the instruction.

The Intelledex robot is much like a microcomputer, with extensive input/output capability that controls a robotic arm. The participant in Intelledex Training learns how to program the robot's microcomputer using Robot BASIC, an extension of Microsoft BASIC. The trainee must be familiar with microcomputers and the BASIC programming language before training is begun. Classes are conducted by the Intelledex Factory Training Department and/or the Regional Training Instructor at the regional center.

On-site training consists of learning at a factory where the participant is or will be employed. The customer firm makes available its own Intelledex robot for the training period. Experience has shown that training is more effective when the trainees are away from their facility and not subject to interruptions from their jobs.

The training is divided into several modules. Module 1 is a one-day course which includes general operation of the robot. It is an introduction to Intelledex robotics. The course is designed to train robot operators, managers, programmers, applications engineers, and service engineers. Topics covered include information about Intelledex (organization and tour of facilities) and an introduction to the robot and vision. The participant studies parts of the robot, modes of operation, system software, safety, and a robot demonstration. Participants study "end effector" design of pneumatic and electrical tools, and sensory input. System operation covers the robot's memory map and power systems.

Module 2 consists of 3 1/2 days of lecture and laboratory, including in-depth robot programming and robot applications. This section includes hands on robot training. Module 2 is for programmers, application engineers, and service engineers. Topics covered include personal computer operation, coordinate systems, manual mode, and standalone mode. Participants study Robot BASIC/Vision, integrated vision, off-line programming, interfacing, application programming (flowcharting, coding, and debugging), hands-on programming exercises, and a callibration overview. Trainees enter Intelledex's Field Support Program, where they learn diagnostics and module/board exchange techniques.

Module 3 trains the participant to service the Module Exchange Level. The five-day course is designed for companies who will do their own servicing of the robot. The program is targeted at service engineers, the persons who will repair the robot. Topics covered include system overview, electrical system, vision system, installation, troubleshooting procedures, mechanical system, and preventive maintenance. Trainees learn such skills as parts identification, callibration, connection of major assemblies, computer vision, service and self-check diagnostics, disassembly and reassembly, alignment, and adjustments. In preventive maintenance, the trainees learn how to service the air system, gearboxes, and electronic cabinet.

Workbooks, manuals, reference guides, and other necessary materials are given to the student. Self-examinations (with answers) are included so that the student can evaluate his or her own progress. Primer materials are mailed in advance of the training to familiarize the student with general terminology. Additional instruction is available from Intelledex Application Engineering to any student who wants to learn Intelledex robotics. Tutoring fees are $35 per hour or $280 per day. A familiarity with BASIC programming language is required before entering training classes. BASIC may be learned at almost any community college, adult education program, correspondence course, vocational class, or self-learned from manuals purchased in a computer software store. Intelledex must have a company's purchase order number before any prospective employee, intern, or worker begins training. However, a student may take the training course before his or her employer purchases an Intelledex robot.

Offered At:
Intelledex Inc.
33840 Eastgate Circle
Corvallis, OR 97333
(503) 758-4700

Time of Course:
One, three-and-a-half, or five days. Class hours are from 8:30 a.m. to 5:00 p.m.

Tuition:
Costs vary from $100 to $600, depending upon length of course and whether it is

	given at the Intelledex Factory or Regional Center, or at the customer's facility.
Robots:	Intelledex robots are used.
Certificate:	A letter of completion is awarded upon satisfactory completion of the training. All necessary materials are given to the student. Workbooks are structured to be used as reference materials.
Additional Services:	Intelledex will make hotel reservations for trainees coming to factory training in Corvallis or to training in Intelledex Regional Centers in Santa Clara, California; Boston, Massachusetts; Minneapolis, Minnesota; or Dallas, Texas. Courses outside of the customer's facility offer lunch and refreshments for the trainees.

EXPERT AUTOMATION/KUKA ROBOTICS TRAINING COURSES

Expert Automation/KUKA Inc. believes that its robotic equipment provides maximum reliability to meet today's varying production requirements. To obtain maximum effectiveness from this equipment requires personnel with proper training. KUKA offers a comprehensive, hands-on structural set of training programs to produce skilled personnel with an understanding of the operation, maintenance, and capabilities of the KUKA line of industrial robots. The courses are responsive to customer needs and are designed to give trainees confidence in their achievements. Admission is based on space availability. The office of Expert Automation/KUKA Inc. should be contacted for changing course schedules.

The first course offered is Programming and Operations. The participants are operators, maintenance technicians, electricians, and engineers. A prerequisite is experience with industrial machinery.

The course description consists of an introduction to the industrial robot, construction of the robot, drive, integrated circuits, and feedback system. Operation of the robot includes the operating elements and application of programs, locking devices, and safety installations.

Programming theory and practice is covered, along with routine maintenance and tracing of errors. Identification of problem cause and repair is covered. The objective of this course is to enable the participants to operate a robot, to render user programs, to correct programs, to do routine maintenance, and to clear simple problems.

The Electrical Maintenance course requires knowledge of digital technique and interpretation of logic circuits, as well as knowledge of digital/analog controls. This course includes introduction to the industrial robot, construction of the robot, specifications, gears, integrated circuit, feedback systems, mechanical construction, electrical construction, and electrical control. Documentation is included, along with operating elements, operating cycle, logic description, interface description, tracing of errors, and hands-on training with the robot. The objective of this course is to enable the participant to identify hardware faults and to clear them.

Mechanical Maintenance is the core of a course to train robot repairers (mechanics). The prerequisite is experience with industrial machines. The course consists of an introduction, the study of construction and drive, operation, and practice maintenance of the robot. Upon completion of this course, participants will be capable of performing normal maintenance and mechanical repairs and calibrations.

Offered At:	Expert Automation/KUKA Inc.
	40675 Mound Road
	Sterling Heights, MI 48078
	(313) 977-0100
Time of Course:	Programming and Operations, one week. Mechanical Maintenance, one week.Electrical Maintenance, two weeks.
Tuition:	$850 for one-week to $1700 for two-week sessions.
Robots:	All training is on KUKA robots. Students interested in obtaining the model numbers of the KUKA line of industrial robots should write directly to Expert Automation/KUKA.
Certificate:	A certificate of completion is awarded.
Additional Services:	In-plant training courses can be arranged upon request. Expert Automation/KUKA can assist in lodging accommodations upon request.

LENNOX EDUCATION PRODUCTS

Lennox Industries, Inc., maintains a division called Lennox Education Products, which has introduced one of the first work-oriented robotics training programs in the nation, a program designed to train robotic application technicians. Lennox Education Products Division pioneered training for the heating and air conditioning industry. Currently, Lennox has introduced a robotics training package priced at $2700 called the Lennox Teaching Robot.

The Lennox course will be marketed to vocational/technical schools, community colleges, major manufacturers, and others interested in starting robotics training programs.

It is particularly useful to women's groups that need to start a high-technology training program for women of varied backgrounds and experience, and for groups seeking to train previously untapped labor sources.

The demand for robotics personnel depends on the fact that few robots can be applied to work situations without modification and adaptation. Jobs currently lie with the robot builders, but there should soon be a shift to the robot users, where the masses of robotics jobs should be. The Lennox Robotics Training Program prepares participants to select, program, and assemble robots. Students are also taught to troubleshoot robotic installation and to maintain robots in industry.

The robotics training programming can be purchased as a package or as individual components customized to specific needs. Persons purchasing the course are sent instructional materials, including a robot and control modules. The package also can be used for instruction in troubleshooting and maintaining robotic applications. It is part of a complete training system with individual components which can be tailored to specific robotic teaching needs.

Each Lennox Teaching Robot contains a single-arm pick-and-place pneumatic manipulator, a valve interface, a Timex/Sinclair 1000 computer and printer, a control panel, a television receiver, and a cassette recorder. The computer, panel, and television are used to establish the robot's programs and the cassette recorder is used to store them.

Instructional materials include manuals, video cassettes, and overhead transparencies. The robotics equipment available include single-arm, double-arm, box-arm, dc servo, and hydraulic servo robots. Other equipment includes manual control switch boxes, programmable controller trainers, and industrial or process control computers.

When offered by a community college, the course will run two semesters. However, for concentrated in-house industrial seminars, the same course can be completed in four one-week intensive sessions.

Offered At:	Lennox Education Products P.O. Box 809000 Dallas, TX 75380 (214) 233-9407
Time of Course:	Four weeks to two semesters
Tuition:	$2700, or cost of community college or in-house seminar program.
Robots:	Lennox Teaching Robot
Certificate:	Certification is issued through the participating community college or in-house seminar.
Additional Services:	Correspondence/home study can be arranged by contacting Lennox Education Products.

NASHVILLE STATE TECHNICAL INSTITUTE

Nashville Tech offers its Automation-Robotics Technology curriculum to provide the student with a broad range of technical skills in the electronic, electrical, and mechanical areas as they relate to the automatic control of manufacturing or other complex systems. This training program includes such topics as microprocessors, transducers, sensing devices, programmable controllers, and hydraulic and pneumatic systems. These individual topics are unified in courses dealing with the application of robots and troubleshooting. The courses provide participants with an understanding of the interaction of these systems as they control complex systems and with hands-on experience in correcting malfunctions in such systems.

There is a broad range of businesses and industries that utilizes automated systems in their operations. Consequently, the number of jobs and their titles are varied and offer excellent opportunities for the graduates of this program. Typical jobs for graduates are:

Maintenance Technician	Responsible for the repair and maintenance of automated manufacturing systems.
Robotic Service Technician	Responsible for maintenance and repair of robots.
Installation Technician	Responsible for the installation and start up of automated systems.

In order to earn the Associate degree (A.S.), the student in this two-year program must complete a minimum of 90 quarter hours of courses. A student without a high school diploma may be admitted into the certificate programs, but he or she must obtain the equivalent of a high school diploma before finishing the program. Courses from other accredited schools are transferable. For an Associate degree, the last 30 credit hours preceding graduation must be completed at Nashville State Technical Institute.

The first year in the Automation-Robotics Technology curriculum is divided into three quarters. In the first quarter the student takes Engineering Drawing, Mechanical Technology Orientation, Introduction to Electrical and Electronic Engineering Technology, Algebra and Trigonometry I, and Effective Writing. In the winter quarter the participant moves on to BASIC computer programming, Electrical Technology I, Algebra and Trigonometry II, and Report Writing. The spring quarter concludes the first year of study with courses in Integrated Circuits, Electrical Technology II, Mechanics, and Speech.

The second year of study begins with Hydraulics and Pneumatics, Electrical Machinery II, Logic Circuits and Boolean Algebra, and Human Relations of Business in the fall quarter. The winter quarter offers Robotics Applications, Microprocessor Principles I, Servomechanisms Systems, and Industrial Instrumentation. In the second, final spring quarter the participant studies Systems Troubleshooting; The Physics of Heat, Light, and Sound; Microprocessor Principles II; Machine Elements II; and an elective.

Nashville Tech offers a cooperative option. The school training program recommends that students selecting the robotics major consider a full-time cooperative work experience between the first and second year during the summer and parallel cooperative work experiences during the second year. The program requires that students understand the basics of robotics from three different fields of study. Two or more parallel work experiences may be required in order to replace a course. Up to 15 hours may be replaced by approved cooperative work experiences.

Offered At:	Nashville State Technical Institute 120 White Bridge Road Nashville, TN 37209 (615) 741-1268
Time of Course:	The Automation-Robotics Technology curriculum generally takes two years to complete. The program usually requires 90 to 99 quarter hours of study.
Tuition:	All full-time students, whether in-state, out-of-state, or foreign national, are charged a $154 maintenance fee per quarter. Out-of-state students will be charged an additional $588 tuition per quarter. Students registered for 12 or more college credit hours are subject to this fee for the entire course load. All part-time students are charged $13 per college credit hour or continuing education unit. Out-of-state students pay an additional tuition of $49 per quarter hour.
Robots:	Cooperative work experience courses allow the student to obtain hands-on experience with a variety of industrial robots at many different work-sites. The students work off-campus for various industries and receive course credit.
Certificate:	Nashville Tech grants the Associate of Science Degree in Automation-Robotics Technology.
Additional Services:	It is Nashville Tech's belief that no qualified student should be denied the opportunity of an education because of financial need. Grants, loans, scholarships, and part-time work opportunities exist. The following financial aid programs can be combined and packaged for individual students: Pell Gran, College Work-Study Program, Supplemental Educational Opportunity Grant, Nashville

Tech Work Scholarship, Guaranteed Student Loan Program, Tennessee Student Assistance Award, and veterans assistance. Persons may write to the Financial Aid Department at Nashville State Technical Institute for further information.

SCHOOLCRAFT COLLEGE

Schoolcraft College, a community college located in Livonia, Michigan, began planning its two robotics programs in January, 1983, by assembling an active advisory committee. Program development centered on the electrical/electronic aspect of robotics. Two unique program offerings are listed in the school's catalogue.

The school finds, as industry has projected, that the school's graduates will be employed based on their electrical maintenance skills or their ability to recommend robotics application to the manufacturing process. The school has been fortunate in having its faculty serve industry internships at Ford Motor Company's Robotics and Automation Applications Consulting Center (RAACE) on a semester-by-semester basis. They are employed by Schoolcraft College, but their work assignments are under the direction of Ford. From all indications, they have become valuable assets to the center by performing applications evaluations, programming, troubleshooting, and instructing Ford personnel and others in robotic fundamentals.

To date, Schoolcraft College has acquired twelve Z-100 Heathkit/Zenith computers with monitors, two Z-25 Heathkit/Zenith printers, six RHINO XR-1 classroom mechanical robots, six HERO-1 programmable electronic robots, and the ASEA, BINKS, and Motion Mate robots. This selection of machines allows the students not only to have access to the introductory courses, but to interact with the electronics and mechanisms of industrial robots.

The school also is teaching principles of CAM (computer-aided manufacturing), a course in which the students are actively involved in writing their own programs for the HEROs and RHINOs, and are assigned projects in order to solve robot application problems. Students are also assigned the role of trouble-shooting and repair work typical in utilizing educational units.

Two majors are offered for the Associate degree, the Robot Application Technician program and the Robot Service Technologist program. A Robot Application Technician seeks ways in which to utilize robots to improve the company's productivity, quality, and to save the company money. The technician is a "behind-the-scenes" member of the team, solving problems and seeking more timely and better ways of performing work. Applications technicians have close ties with management and process engineers.

The two-year program allows the first-year student to study basic machining, industrial management, technical math, materials science, manufacturing theory, machine operation, robotics, and BASIC computer programming. In the second year the student studies manufacturing systems, statistics and quality control, engineering drawing, physical metallurgy, product analysis, cost estimating, mechanical/magnetic principles, technical writing, hydraulics and pneumatics, and manufacturing methods and evaluation.

The other robotics major is the Robot Service Technologist program. At the end of three years the Associate degree is awarded. The job of the robot technologist is to implement the ideas of the engineer and improve them when applicable. The technologist serves as a liaison between the engineer and the robot application specialist. Technologists engage in hands-on work with the robot, such as setting up the system and going out on service calls for maintenance and repair of robots.

Subjects studied in the service program include dc fundamentals, electronics math, BASIC programming, ac fundamentals, diodes and transistors, technical writing, digital circuits, amplifiers, instrumentation/equipment certification, introduction to robotics, digital logic circuits, introduction to microprocessors, linear circuits, CAM, mechanical/magnetic principles, electronic data communications, microcomputer interfaces, industrial controls, robotic systems, human performance technology, and robotic system analysis and troubleshooting.

Offered At:	Schoolcraft College 18600 Haggerty Road Livonia, MI 48152 (313) 591-6400
Time of Course:	2-3 years of study is required.
Tuition:	Resident tuition is $25.50 per credit hour. Non resident tuition is $34.50 per credit hour. Out-of State and Foreign student tuition is $51.50 per credit hour.
Robots:	The school maintains six RHINO XR-1 classroom mechanical robots, six HERO-1 programmable electronic robots, and the ASEA, BINKS, and Motion Mate robots.
Certificate:	The Associate degree is offered after two to three years of study, depending on the major.
Additional Services:	Students obtain hands-on experience with industrial robots. An advisory committee serves industry internships at the Robotics and Automation Applications Consulting Center at Ford Motor Company on a semester-by-semester basis. Faculty performs applications evaluations, programming and trouble-shooting. In addition, it instructs Ford personnel and others in robotics and teaches students in the community college classes.

STAPLES TECHNICAL INSTITUTE

Staples Technical Institute of Staples, Minnesota, offers 22 months of intensive robotics technician training. The course design provides a bond between the Staples robotics program and the needs of industry. Hands-on training includes conducting feasibility studies, finding suitable robots, programming robots for specific work, and designing accessories and specialized tools.

Instruction is given in installing and maintaining the robots for field projects. Subjects studied include technical mathematics, electronics, technical robotics writing, CNC operation, welding, hydraulics, and pneumatics.

STI students are available to companies who wish to conduct research with robots to develop applications. Installation and maintenance can be executed by students. Students can complete feasibility studies for businesses to determine how robots can best be used in a particular firm.

Students are trained to think in terms of creating complete packages for companies and developing real applications, instead of working only with theoretical projects. Local manufacturers have had input into the course design since 1981. Free equipment is often donated to the school by manufacturers. As part of agreements, manufacturers often donate robots in exchange for students modifying the equipment to perform functions requested by the manufacturers and other industries.

Students design work cells which perform various applications. Participants work to find robots that perform applications within various company's specifications. They write to manufacturers and perform research, as well as work hands-on in modifying the robots. For further information, interested persons can write directly to the school at this address:

Staples Technical Institutes
419 N.E. 3rd Street
Staples, MN 56479
(218) 894-2430

UNIVERSITY OF SOUTHERN CALIFORNIA GRADUATE STUDY IN ROBOTICS

The University of Southern California (USC) maintains a Robotics Institute for graduate studies. There is no formal degree program in robotics *per se,* but the college offers many courses and opportunities for study and research in robotics. Students apply to the academic department of their choice; the academic program must conform to the degree program of a department within the School of Engineering. Most students interested in robotics take Introduction to Robotics and Basic Robotics Laboratory.

The Robotics Research Laboratory currently is equipped with a Unimation PUMA robot, an IBM 7565 manipulator, and a Digital Equipment Corporation VAX-11/750 computer. This laboratory will also house a special 3-degrees-of-freedom arm being constructed at USC to support experiments in robot control. The Robotics Teaching Laboratory currently is equipped with two Microbot MiniMower-5 arms, each controlled by an Apple-II computer. The laboratory is used in conjunction with the Basic Robotics Laboratory course and is also available for special student projects.

The Image Processing Laboratory has been involved since 1971 in basic and applied studies in the fields of digital and optical methods of image processing. Facilities include PDP-11/40 and PDP-11/55 minicomputers, a precision scanning microdensitometer, a high-resolution color film plotter, real-time digital video image display systems, lasers, and associated optical equipment. The Production Systems Laboratory houses an automated mini-warehouse (300 storage positions), an NC lathe and mill, and a USC-built 4-degrees-of-freedom fixed-stop robot. The Human Factors Laboratory offers equipment for work-measurement studies and other industrial engineering experiments.

The Computer Research Laboratory provides Computer Engineering faculty and students with facilities for advanced research in all aspects of computer research. Centered on a DEC VAX-11/750 superminicomputer, the system includes peripherals for high-quality printed output, color graphics input, and a network of SUN workstations. The Engineering Computer Laboratory manages six DEC-20 mainframe computers, a host of VAX superminicomputers, high-quality output devices, and several hundred terminals across the campus.

An introductory course in robotics offers the fundamental concepts of kinematics and actuators. Architecture of robot systems, dynamics, force and control, and sensing by vision are studied. Such topics as proximity, touch, robot programming languages, planning and modelling are explored.

The Basic Robotics Laboratory course offers laboratory exercises using microcomputers and small robots involving software simulation, positioning, collision avoidance, and sensor interfaces. Software Methods in Robotics offers numerous robot programming languages, robot architectures and operating systems, design of software interfaces, geometric modelling and simulation, grasping, and planning of robot tasks.

A course in Analytical Methods in Robotics emphasizes robot manipulators, dynamics, and control. The Robotics Institute offers courses in pattern recognition, computer vision, and image processing. Participants study mathematical pattern recognition, machine perception, digital signal processing, and speech signal recognition. Courses in artificial intelligence explore problem-solving techniques in robotics. Such topics as description and recognition of objects by robots, shape analysis, edge and region segmentation, texture, knowledge-based systems, and image understanding are explored in the Machine Perception course.

A special emphasis in manufacturing offers courses in computer-aided design, computer-aided manufacturing, engineering economy, and human factors in engineering design. The program offers an opportunity for research in designing person-machine systems.

A variety of robotics research areas are offered at USC. Projects share facilities and expertise, and each is managed by a faculty member. Current robot research projects at the Robotics Institute include the design of languages for robot programming. A course in design and construction of a 3-link robot for experiments in control algorithms (methods/techniques) is offered. One research project involves the conversion of a prosthetic arm from analog to digital control. New methods for estimating robot costs and benefits currently is being researched. The design of a new teleoperator controller using force feedback is being worked out. Design of novel pick-and-place manipulators to achieve lower cost than those currently available is being researched. Other current research projects include experiments with knowledge systems based on organization modeling, image-understanding systems, and expert systems for robot kinematics and multi-arm cooperation.

Offered At:	Robotics Institute University of Southern California Los Angeles, CA 90089 (213) 743-2311
Time of Course:	Open-ended. Most courses are based on the semester system.
Tuition:	A limited number of research assistantships are available through the Robotics Institute. Students should write to the Director of the Robots Institute for an application. Courses can also be taken at the regular USC graduate tuition fees. The fees may change annually, and students are asked to write to USC for current tuition rates.
Robots:	The Robotics Research Laboratory is equipped with a Unimation PUMA robot, an IBM 7565 manipulator, and a DEC VAX-11/750 computer. This laboratory will also house a special 3-DOF arm being constructed at USC to support experiments in robot control. Other robotic equipment include two Microbot MiniMover-5 arms, and various optical equipment, lasers, plotters, and computers.
Certificate:	There is no formal degree program in robotics, but USC offers many opportunities for study and research in robotics at the Robotics Institute.

Appendix A

The Robotics Degree Programs

By the end of 1983 there were 27 schools offering full degree programs in robotics. The midwest topped the nation with 10 programs. Two-year community colleges offered 12 out of the 27 robotics degree programs with preparation centered on robotics technology and maintenance courses.

The majority of equipment in robotics education training programs consists of small table-top arms. A 1983 survey by Robotics International of the Society of Manufacturing Engineers revealed that most of the equipment used in training is centered in four-year colleges. The largest number of robots is also found in four-year colleges. This is in contrast to the fact that the largest number of robot degree programs is found at the community college (two-year) level. Seiko and Unimation robots are used most frequently in these programs.

In the United States today there were only 4,100 robots in 1980. By the end of this decade that figure is projected to leap to 21,575 robots. Let's take a look at a list of schools surveyed by the Society of Manufacturing Engineers at the end of 1983 to see who else is offering courses and degrees in robotics. (A more recent directory will be out as this book goes to press and is available free from SME). Interested persons should send a large self-addressed stamped envelope to Robotics International, Society of Manufacturing Engineers, One SME Drive, P.O. Box 930, Dearborn, Michigan 48121. The following list of schools is reprinted with the permission of the Society of Manufacturing Engineers.

TWO-YEAR ROBOT DEGREE PROGRAMS

COLORADO

Community College of Denver - Red Rocks
Science and Technology
12600 W. 6th Avenue
Golden, CO 80401
Robotics Contact Person: P.E. Perkins
                                   L. E. Deaver
Phone: (303) 988-6160, Ext. 388
Program Title: Flexible Automation - Robotics
Courses Available: Robotics, Hydraulics, Pneumatics, Microprocessors, Maintenance/Repair, Logics
Length of Program/Course: 2 years
Enrollment Requirements: Acceptance to program
Robot Laboratory Available: At a local manufacturing facility

Degree/Certification Awarded: A.A.S.

FLORIDA

Broward Community College
Engineering Technology
3501 S.W. Davie Road
Ft. Lauderdale, FL 33314

Robotics Contact Person: Dr. Samuel L. Oppenheimer
Phone: (305) 475-6683
Program Title: Robotics/Process Control
Courses Available: Robotics, Hydraulics/Pneumatics, Integrated Systems, Microprocessors, Logics (Prerequisite)
Length of Program/Course: 2 years, 48 weeks
Enrollment Requirements: A.S. in Electronics Technology

Robot Laboratory Available: Under Development
Major Laboratory Equipment: Microprocessor systems, servomechanisms, manipulator.
Degree/Certification Awarded: Institutional Certificate (beyond A.S. Degree)

Gulf Coast Community College
Division-Technology
5230 W. Highway 98
Panama City, FL 32401

Robotics Contact Person: William Schilling
Phone: (904) 769-1551
Program Title: Electronics Technology/Robotic Option
Length of Program/Course: 2 years, 64 weeks
Enrollment Requirements: High School Graduate, Acceptance to Program
Robot Laboratory Available: Yes
Major Laboratory Equipment: Lever, gear & pulley trainers, Electrical Motor Trainers; Hydraulic, Pneumatic and fluidic trainers; industrial control trainers; CAD equipment; microcomputers; xy equipment
Degree/Certification Awarded: A.S.

ILLINOIS

College of Du Page
22nd Street & Lambert Road
Glen Ellyn, IL 60137

Robotics Contact Person: Dr. James B. McCord
Phone: (312) 858-2800

Program Title: Robotics Technology (Proposed)
Courses Available: Hydraulics, Pneumatics, CAD/CAM, N/C, Integrated Systems, Microprocessors, Logics
Length of Program: 2 years
Enrollment Requirements: High School Graduate

MICHIGAN

Henry Ford Community College
Career and Occupational Studies
5101 Evergreen Road
Dearborn, MI 48128

Robotics Contact Person: Mr. John Nagohosian
Phone: (313) 271-0445
Program Title: Electrical/Electronics Technology - Robotics/Automation Option
Courses Available: Robotics, Hydraulics, Pneumatics, N/C, Integrated Systems, Microprocessors, Maintenance/Repair, Logics
Length of Program/Course: 2 years, 16 weeks each course
Enrollment Requirements: High School Graduate, Acceptance to Program
Robot Laboratory Available: Under Development
Major Laboratory Equipment: Pneumatic, hydraulic, electronic equipment of a wide variety
Degree/Certification Awarded: A.S.

Macomb Community College
Mechanical Technology
14500 Twelve Mile Road
Warren, MI 48093

Robotics Contact Person: Laurence Ford
Phone: (313) 445-7455
Program Title: Robotics Technology
Courses Available: Robotics, Hydraulics, Pneumatics, CAD/CAM, N/C, Integrated Systems, Microprocessors, Maintenance/Repair, Logics
Length of Program/Course: 2 years, 16 weeks
Enrollment Requirements: Acceptance to program
Robot Laboratory Available: Yes
Major Laboratory Equipment: Unimate, Seiko, Copperweld, ASEA.
Degree/Certification Awarded: A.S.

C.S. Mott Community College
1401 East Court Street
Flint, MI 48503

Robotics Contact Person: John Ortiz
Phone: (313) 762-0387
Program Title: Robotics
Courses Available: Hydraulics, Pneumatics, CAD/CAM, N/C, Integrated Systems, Microprocessors, Maintenance/Repair, Logics

Oakland Community College
Applied Technology Department
2900 Featherstone Road

Auburn Heights, MI 48016

Robotics Contact Person: Dr. Bill J. Rose
                           Edward Konopka
Phone: (313) 852-1000. Ext. 306
Program Title: Robotics Technology
Courses Available: Robotics, Hydraulics, Pneumatics, CAD/CAM (Development Stage), N/C, Integrated Systems, Microprocessors, Maintenance/Repair, Logics
Length of Program/Course: 2 years, 14 weeks
Enrollment Requirements: High School Graduate, Acceptance to Program
Robot Laboratory Available: Under Development
Degree/Certification Awarded: A.S.

Schoolcraft College
Technology Department
18600 Haggerty Road
Livonia, MI 48152

Robotics Contact Person: Fernon P. Feenstra
Phone: (313) 591-6400, Ext. 530
Program Title: Robot Technician and Technologist
Courses Available: Robotics, Hydraulics, Pneumatics, N/C, Microprocessors, Maintenance/Repair
Length of Program/Course: 2 years, 16 weeks each course
Enrollment Requirements: High School Graduate, Manufacturing Experience
Robot Laboratory Available: Under Development
Degree/Certification Awarded: A.S.

Washtenaw Community College
P.O. Box D-1
Ann Arbor, MI 48106

Robotics Contact Person: George Agin
Phone: (313) 973-3474
Program Title: Robotics
Courses Available: Robotics, Hydraulics, Pneumatics, N/C, Microprocessors, Maintenance/Repair

Length of Program/Course: 2 years, 15 1/2 weeks
Enrollment Requirements: High School Graduate, Must have College Level Algebra Competence
Robot Laboratory Available: Under Development
Major Laboratory Equipment: Under Development
Degree/Certification Awarded: A.S.

OHIO

University of Toledo
Community & Technical College
Engineering Technology
Toledo, OH 43606

Robotics Contact Person: Dr. C. Ziegler
Phone: (419) 537-3163
Program Title: Industrial Engineering Technology
Courses Available: Hydraulics, Pneumatics, N/C, Microprocessors, Logics

SOUTH CAROLINA

Piedmont Technical College
Electronic Engineering Technology
P.O. Drawer 1467
Greenwood, SC 29648

Robotics Contact Person: Mr. James Rehg
Phone: (803) 223-8357
Program Title: Robotics Technology
Courses Available: Robotics, Hydraulics, Pneumatics, Integrated Systems, Microprocessors, Maintenance/Repair, Logics
Length of Program/Course: 2 years, 11 weeks
Enrollment Requirements: Acceptance to Program
Robot Laboratory Available: Yes
Major Laboratory Equipment: Cincinnati Milacron T3, PUMA 260, Seiko 700, RHINO, computers, programmable controllers
Degree/Certification Awarded: A.S.

FOUR-YEAR ROBOT DEGREE AND OPTION PROGRAMS

FLORIDA

University of Miami
Industrial Engineering Department
Coral Gables, FL 33124

Robotics Contact Person: Professor Carl M. Kromp
Phone: (305) 284-4040
Program Title: Manufacturing Engineering Option, B.S.I.E. Program

Courses Available: Robotics, Hydraulics, CAM, N/C,
Microprocessors
Length of Program/Course: 4 years, 14 weeks
Robot Laboratory Available: Under Development
Major Laboratory Equipment: Seiko 700 Robot with TI
Controller (510)
Degree/Certification Awarded: B.S.I.E.—Manufacturing
Engineering Option

University of Florida
Industrial and Systems Engineering
Weil Hall
Gainesville, FL 32611

Robotics Contact Person: Thomas Kisko
Phone: (904) 392-1464
Program Title: Industrial and Systems Engineering
Courses Available: Robotics, Hydraulics, Pneumatics,
CAD/CAM, Integrated Systems,
Microprocessors, Logics
Length of Program/Course: 4 years, 15 weeks
Enrollment Requirements: Acceptance to Program
Robot Laboratory Available: Yes
Major Laboratory Equipment: 2 MiniMovers, PDP 11/34,
Apple Computers, SYM's,
Fischer-Technik Equip-
ment

Degree/Certification Awarded: M.S., B.S.

Lawrence Institute of Technology
Mechanical Engineering Department
21000 West 10 Mile Road
Southfield, MI 48075

Robotics Contact Person: Wayne M. Brehob
Phone: (313) 356-0200
Program Title: Mechanical Engineering, Manufacturing
Option
Courses Available: Robotics, Hydraulics, CAD/CAM,
Microprocessors.
Length of Program/Course: 4 years, 10 weeks days, 7
weeks evenings
Enrollment Requirements: Acceptance to Program
Robot Laboratory Available: Under Development
Major Laboratory Equipment: Gerber CAD, Bridgeport
NC, RHINO

State University of New York
College of Technology
811 Court Street
Utica, NY 13502

Robotics Contact Person: Professor Atlas Hsie
Phone: (315) 792-3505
Program Title: Industrial Manufacturing, Robotics
Courses Available: Robotics, Applications, Robotics and
CAM, Robotics Seminar, Computer
Vision for Robots, Robot Mechanism,
Robot Sensors and Interfacing, Ro-
bot Control Systems, Hydraulics,
Pneumatics, CAD/CAM, N/C, In-
tegrated Systems, Logics
Length of Program/Course: 4 years, 14 weeks
Enrollment Requirements: Associate Degree
Robot Laboratory Available: Yes
Major Laboratory Equipment: PUMA 600, PUMA 250,
Seiko, RHINO, Mini-
Movers, Computers PC's,
etc.
Degree/Certification Awarded: B.S.

Duke University
Electrical Engineering
Durham, NC 27706

Robotics Contact Person: Paul P. Wang
Phone: (919) 684-3123
Program Title: Machine Intelligence and Robotics
Courses Available: Robotics, CAD/CAM, Integrated
Systems, Microprocessors, Logics
Length of Program/Course: 4 years, 14 weeks
Robot Laboratory Available: Yes
Major Laboratory Equipment: Micro-mover-5, IRI Robots
Degree/Certification Awarded: B.S.

Oregon State University
Industrial Engineering
Corvallis, OR 97331

Robotics Contact Person: Dr. Gene Fichter
Phone: (503) 754-4505

Program Title: Manufacturing Engineering Co-Op
Program
Courses Available: Robotics, CAD/CAM, N/C
Length of Program/Course: 5 years, 9 weeks
Enrollment Requirements: Acceptance to Program
Robot Laboratory Available: Under Development

PENNSYLVANIA

Carnegie-Mellon University
Robotics Institute
Pittsburgh, PA 15213

Robotics Contact Person: Paul K. Wright
Phone: (412) 578-3529
Program Title: Manufacturing Science and Robotics
Courses Available: Robotics, CAD/CAM, Microprocessors
Length of Program/Course: 4 years, 14 weeks
Enrollment Requirements: Acceptance to Program
Robot Laboratory Available: Yes
Major Laboratory Equipment: Cincinnati Milacron,
Fanuc, PUMA and Seiko
Robots, Sensor Equipment, etc.
Degree/Certification Awarded: B.M.E.

SOUTH CAROLINA

University of South Carolina
Electrical and Computer Engineering
Columbia, SC 29208

Robotics Contact Person: Ronald D. Bonnell
Phone: (803) 777-3075
Program Title: Machine Intelligence
Courses Available: Robotics, Computer Graphics (CAD),
Microprocessors, Logics
Length of Program/Course: 4 years, 14 weeks
Enrollment Requirements: Undergraduate Degree in
Engineering (ABET)
Robot Laboratory Available: Under Development.
Operate a Machine Intelligence Lab, Computer
Graphics Lab, Microprocessor Lab, and Computer
Interfacing Lab
Major Laboratory Equipment: VAX 11/780; Comtal Vision System, Speech
Systems; 2 Megatek
Graphics Work Stations; 8
NC development systems
Degree/Certification Awarded: B.S., M.S. (ECE Specialty in Machine Intelligence)

GRADUATE ROBOT PROGRAMS

GEORGIA

Georgia Institute of Technology
Mechanical Engineering
Atlanta, GA 30332

Robotics Contact Person: Stephen L. Dickerson
Phone: (404) 894-3255

Program Title: Programmable Automation
Courses Available: Robotics, Hydraulics, Pneumatics,
CAD/CAM, N/C, Integrated
Systems, Microprocessors, Logics
Enrollment Requirements: Undergraduate Degree
Robot Laboratory Available: Yes
Major Laboratory Equipment: Cincinnati Milacron
T3, Terak 8600 color
graphics system
Degree/Certification Awarded: M.S.

ILLINOIS

University of Illinois at Chicago
College of Engineering
Box 4348
Chicago, IL 60680

Robotics Contact Person: Dr. Paul M. Chung
Phone: (312) 996-2400
Program Title: Electrical Engineering or Mechanical
Engineering with Robotics Concentration
Courses Available: Robotics, CAD/CAM, Integrated
Systems, Microprocessors, Logics
Length of Program/Course: 2 years/4 years
1 1/2 year—M.S., 4 yr
—Ph.D.
Enrollment Requirements: Undergraduate Degree
Robot Laboratory Available: Yes
Degree/Certification Awarded: M.S., Ph.D.

NEW YORK

Rensselaer Polytechnic Institute
Electrical Computer & Systems Engineering
Troy, NY 12181

Robotics Contact Person: George N. Saridis
Phone: (518) 270-6316
Program Title: Robotics and Automation
Courses Available: Robotics, CAD/CAM, Micro-
processors, Logics
Enrollment Requirements: Acceptance to Department
Robot Laboratory Available: Yes
Major Laboratory Equipment: Cincinnati Milacron T3,
Hydraulic Cylinder Ro-
bot, Pneumatic robot,
MiniMover

NORTH CAROLINA

Duke University
Electrical Engineering Department
Durham, NC 27706

Robotics Contact Person: Paul P. Wang
Phone: (919) 684-3123
Program Title: Machine Intelligence and Robotics
Courses Available: Robotics, CAD/CAM, Integrated
Systems, Microprocessors, Logics
Enrollment Requirements: Undergraduate Degree
Robot Laboratory Available: Yes
Major Laboratory Equipment: Micro MiniMover-5, 1 RT
Robot
Degree/Certification Awarded: M.S., Ph.D.

PENNSYLVANIA

Drexel University
Electrical & Computer Engineering
32nd and Chestnut
Philadelphia, PA 19104

Robotics Contact Person: Dr. Richard Klafter
Phone: (215) 895-2223
Program Title: Robotics
Courses Available: Robotics, Microprocessors
Enrollment Requirements: Acceptance to Program,
Undergraduate Degree
Robot Laboratory Available: Yes
Major Laboratory Equipment: PUMA 550, LSI-11 system
Degree/Certification Awarded: M.S.

Carnegie-Mellon University
Robotics Institute
Pittsburgh, PA 15213

Robotics Contact Person: Paul K. Wright
Phone: (412) 578-3529
Program Title: Manufacturing Science and Robotics
Courses Available: Robotics, CAD/CAM, Microprocessors
Enrollment Requirements: Acceptance to Program
Robot Laboratory Available: Yes
Major Laboratory Equipment: Cincinnati Milacron,
Fanuc, PUMA and Seiko
Robots, Sensory Equip-
ment, etc.
Degree/Certification Awarded: M.E.

TEXAS

University of Houston
Industrial Engineering
4800 Calhoun
Houston, TX 77004

Robotics Contact Person: Nelson Marquina
Phone: (713) 749-2543
Program Title: Robotics and Artificial Intelligence
Courses Available: Robotics, CAD/CAM, N/C, Integrated
Systems Microprocessors, Artificial
Intelligence, Decision Support
Systems
Enrollment Requirements: Undergraduate Degree
Robot Laboratory Available: Under Development

TWO-YEAR SCHOOLS OFFERING ROBOT COURSES

ALABAMA

Wallace State Community College
Technical College
P.O. Box 250
Hanceville, AL 35077
Robotics Contact Person: Rayburn Williams
Phone: (205) 352-6403

Program Title: Industrial Electronics
Courses Available: Robotics, Hydraulics, N/C,
Microprocessors, Logics
Length of Program/Course: 2 years, 11 weeks
Enrollment Requirements: High School Graduate
Robot Laboratory Available: Under Development
Degree/Certification Awarded: A.S.

ARIZONA

Yavapai College
1100 E. Sheldon Street
Prescott, AZ 86301

Robotics Contact Person: Larry Strom
Phone: (602) 445-7300
Program Title: Industrial Technology
Courses Available: Robotics, Hydraulics, CAD/CAM, N/C, Microprocessors
Length of Program/Course: 2 years, 18 weeks
Enrollment Requirements: Acceptance to Program
Robot Laboratory Available: Under Development

FLORIDA

Miami-Dade Community College
North Campus, Electronics Dept.
11380 NW 27th Avenue
Miami, FL

Robotics Contact Person: Peter Maitland
Phone: (305) 685-4243
Program Title: Engineering Technology
Courses Available: Robotics, Hydraulics, Microprocessors, Logics
Length of Program/Course: 2 years, 16 weeks
Enrollment Requirements: Acceptance to Program
Robot Laboratory Available: Under Development
Major Laboratory Equipment: Microcomputers
Degree/Certification Awarded: A.S.

GEORGIA

Dekalb College
Electronics Dept.
495 Indian Creek Drive
Clarkston, GA 30021

Robotics Contact Person: Kenneth E. Kent
Phone: (404) 292-1520
Program Title: Electromechanical/Engineering Technology
Courses Available: Robotics, Hydraulics, Pneumatics, CAD/CAM, N/C, Microprocessors, Maintenance/Repair, Logics
Length of Program/Course: 2 years
Enrollment Requirements: High School Graduate, Acceptance to Program
Robot Laboratory Available: No
Degree/Certification Awarded: A.S.

ILLINOIS

Carl Sandburg College
Galesburg, IL 61401
Robotics Contact Person: Don Crist
Phone: (309) 344-2518
Courses Available: Robotics, Hydraulics, Pneumatics, CAD/CAM, N/C
Length of Program/Course: 2 years, 10 weeks
Enrollment Requirements: High School Graduate
Robot Laboratory Available: No

Moraine Valley Community College
10900 S. 88th Avenue
Palos Hills, IL 60465

Robotics Contact Person: Robert G. Backstrom
Phone: (312) 974-4300
Courses Available: Robotics, Hydraulics, Pneumatics, Microprocessors
Length of Program/Course: 2 years
Enrollment Requirements: Acceptance to Program
Robot Laboratory Available: Under Development

Triton College
2000 Fifth Avenue
River Grove, IL 60171

Robotics Contact Person: Neal Meredith
Phone: (312) 456-0300
Program Title: Machine Technology
Courses Available: Robotics, Hydraulics, Pneumatics, N/C, Microprocessors, Computer Graphics
Length of Program/Course: 2 years, 17 weeks
Enrollment Requirements: High School Graduate
Robot Laboratory Available: Under Development
Degree/Certification Awarded: Robotics Certificate, A.A.S.

INDIANA

Vincennes University
Technology Division
Vincennes, IN 47591

Robotics Contact Person: Dean Eavey
Phone: (812) 885-4447 or 4448
Program Title: Electronics Technology
Courses Available: Robotics, Hydraulics, Pneumatics, CAD/CAM, N/C, Microprocessors, Maintenance/Repair
Length of Program/Course: 2 years, 16 weeks

Enrollment Requirements: High School Graduate
Robot Laboratory Available: Under Development
Degree/Certification Awarded: A.S.

IOWA

Marshalltown Community College
Industrial Technology Dept.
3700 South Center
Marshalltown, IA 50158

Robotics Contact Person: Jeff Dodge
Phone: (515) 752-7106
Program Title: Electronics Technology
Courses Available: Robotics, Hydraulics, Pneumatics, N/C, Microprocessors
Length of Program/Course: 2 years, 18 weeks
Enrollment Requirements: Acceptance to Program
Robot Laboratory Available: Under Development
Major Laboratory Equipment: Commodore Computers, 6800 series Microprocessors, ARMDROID I arm/manipulators.

MARYLAND

Howard Community College
Community Education Division
Little Patuxent Parkway
Columbia, MD 21044

Robotics Contact Person: Anne Hux
Phone: (301) 992-4823
Program Title: Community Education
Courses Available: Intro. to Industrial Robots
Length of Program/Course: 6 weeks
Robot Laboratory Available: No

MICHIGAN

Delta College
Technical Division
University Center, MI 48710

Robotics Contact Person: Don Holzhei
Phone: (517) 686-9442
Program Title: Robot Technology
Courses Available: Robotics, Hydraulics, Pneumatics, N/C, Microprocessors, Logics
Length of Program/Course: 2 years, 10 and 15 weeks
Enrollment Requirements: High School Graduate
Robot Laboratory Available: Under Development
Degree/Certification Awarded: A.S.

Grand Rapids Junior College
Technology Division
143 Bostwick, N.E.
Grand Rapids, MI 49503

Robotics Contact Person: Don Boyer
Phone: (616) 456-4860
Program Title: Technology
Courses Available: Robotics, Hydraulics, Microprocessors, Maintenance/Repair, Logics
Length of Program/Course: 2 years, 16 weeks
Enrollment Requirements: High School Graduate
Robot Laboratory Available: Under Development
Major Laboratory Equipment: Complete hydraulics and electronics lab

St. Clair County Community College
323 Erie Street
Port Huron, MI 48060

Robotics Contact Person: Francis J. Mitchell
Phone: (313) 984-3881
Program Title: Mechanical/Industrial Technology
Courses Available: Robotics, Hydraulics, Pneumatics, CAD/CAM, N/C, Maintenance/Repair
Length of Program/Course: 2 years, 16 weeks
Enrollment Requirements: High School Graduate, Acceptance to Program
Robot Laboratory Available: Under Development
Degree/Certification Awarded: A.S.

NORTH DAKOTA

North Dakota State School of Science
Technical Division
Wahpeton, ND 58075

Robotics Contact Person: Bernard Anderson
Phone: (701) 671-2278
Courses Available: Robotics, CAD/CAM, Microprocessors
Length of Program/Course: 2 years, 12 weeks
Enrollment Requirements: Acceptance to Program
Robot Laboratory Available: Under Development

OHIO

Central Ohio Technical College
Engineering Division
University Drive
Newark, OH 43055

Robotics Contact Person: James E. Goodman
Phone: (614) 366-9250

Program Title: EMT Systems
Courses Available: Robotics, Hydraulics, Pneumatics, CAD/CAM, N/C, Microprocessors, Logics
Length of Program/Course: 2 years, 11 weeks
Enrollment Requirements: High School Graduate, Acceptance to Program
Robot Laboratory Available: Yes
Major Laboratory Equipment: RHINO XR-1, Motorola MEK 680002, Digital PDP-11/70
Degree/Certification Awarded: A.S.

Firelands College, B.G.S.U.
901 Rye Beach Road
Huron, OH 44839

Robotics Contact Person: John Kovalchuck
Phone: (419) 433-5560, Ext. 279
Program Title: Engineering Technology/Applied Sciences
Courses Available: Robotics, Hydraulics and Pneumatics, offered as one course
Length of Program/Course: 2 years, 14 weeks
Enrollment Requirements: High School Graduate
Robot Laboratory Available: No
Degree/Certification Awarded: A.S.

Terra Technical College
Engineering Department
1220 Cedar Street
Fremont, OH 43420

Robotics Contact Person: Gordon Saan
Phone: (419) 334-3886
Courses Available: Robotics, Hydraulics, Pneumatics, Integrated Systems, Microprocessors, Logics
Length of Program/Course: 2 years
Enrollment Requirements: Acceptance to Program
Robot Laboratory Available: Under Development
Major Laboratory Equipment: RHINO Robots, 4 Apple Systems

TENNESSEE

Motlow State Community College
Industrial Technology
Tullahoma, TN 37388

Robotics Contact Person: Jasper Templeton
Phone: (615) 455-8511
Program Title: Industrial Technology
Courses Available: Robotics, N/C, Microprocessors
Length of Program/Course: 2 years, 10 weeks
Enrollment Requirements: High School Graduate
Robot Laboratory Available: No
Degree/Certification Awarded: Apprenticeship Machinist Journeyman's Card

WISCONSIN

Moraine Park Technical Institute
Fond du Lac Campus
235 N. National Avenue
Fond du Lac, WI 54935

Robotics Contact Person: James Yu
Phone: (414) 922-8611
Program Title: Electromechanical Technician
Courses Available: Robotics, Hydraulics, Pneumatics, CAD/CAM, N/C, Integrated Systems, Microprocessors, Maintenance/Repair, Logics
Length of Program/Course: 2 years, 18 weeks
Enrollment Requirements: High School Graduate
Robot Laboratory Available: Under Development
Major Laboratory Equipment: Instructional Robot, Hydraulic Trainers, PC's, Logic Trainers, Linkage Trainers
Degree/Certification Awarded: A.S.

Waukesha County Technical Institute
800 Main Street
Pewaukee, WI 53072

Robotics Contact Person: Ron Eigenschink
Phone: (414) 548-5202
Courses Available: Robotics, Hydraulics, Pneumatics, CAD/CAM, N/C, Integrated Systems, Microprocessors, Maintenance/Repair, Logics
Length of Program/Course: 2 years, 28 weeks
Enrollment Requirements: Acceptance to Program
Robot Laboratory Available: Yes
Degree/Certification Awarded: State Certificate

GRADUATE SCHOOLS OFFERING ROBOT COURSES

CALIFORNIA

University of Southern California
Electrical Engineering
PHE 318, MD-0272
Los Angeles, CA 90089
Robotics Contact Person: Barry Soroka
Phone: (213) 743-5504
Courses Available: Robotics, CAD/CAM,
                     Microprocessors
Length of Program/Course: 4 years, 14 weeks
Enrollment Requirements: Undergraduate Degree
Robot Laboratory Available: Yes
Major Laboratory Equipment: Microbot Mini Mover,
                             Apple II Computers

DELAWARE

University of Delaware
Mechanical and Aerospace Engineering
Newark, DE 19711

Robotics Contact Person: Dr. Alok Kumar
Phone: (302) 738-2889
Program Title: Mechanical Engineering
Courses Available: Robotics, Hydraulics, Pneumatics,
                     CAD/CAM
Enrollment Requirements: Undergraduate Degree
Robot Laboratory Available: Under Development

FLORIDA

University of Florida
Industrial and Systems Engineering
Weil Hall
Gainesville, FL 32611

Robotics Contact Person: Thomas Kisko
Phone: (904) 392-1464
Program Title: Industrial and Systems Engineering
Courses Available: Robotics, Hydraulics, Pneumatics,
                     CAD/CAM, Integrated Systems,
                     Microprocessors, Logics
Enrollment Requirements: Acceptance to Program
Robot Laboratory Available: Yes
Major Laboratory Equipment: 2 Mini Movers, PDP
                             11/34, Apple Com-
                             puter, SYM's, Fischer
                             Technik Equipment
Degree/Certification Awarded: M.S., B.S.

University of Central Florida
Industrial Engineering & Management Systems
P.O. Box 2500
Orlando, FL 32816

Robotics Contact Person: John E. Biegel
Phone: (305) 275-2236
Program Title: Industrial Engineering
Courses Available: Robotics, CAD/CAM,
                     Microprocessors, Logics
Enrollment Requirements: Undergraduate Degree
Robot Laboratory Available: Under Development
Major Laboratory Equipment: PUMA, Microcomputer
                             with graphics

ILLINOIS

Southern Illinois University
College of Engineering and Technology
Carbondale, IL 62901

Robotics Contact Person: Dr. Dale Besterfield
Phone: (618) 453-4321
Program Title: M.S. Program in Engineering
Courses Available: Robotics, N/C
Enrollment Requirements: Acceptance to Program
Robot Laboratory Available: Yes
Major Laboratory Equipment: PUMA, Unimate, 4
Microbots
Degree/Certification Awarded: M.S.

INDIANA

Rose-Hulman Institute of Technology
Mechanical Engineering
5500 Wabash Avenue
Terre Haute, IN 47803

Robotics Contact Person: Dr. Donald G. Morin
Phone: (812) 877-1511
Program Title: Mechanical Engineering
Courses Available: Robotics, CAD/CAM,
                     Microprocessors, Logics
Enrollment Requirements: Acceptance to Program
Robot Laboratory Available: Yes
Major Laboratory Equipment: RHINO robot, several
                             computers
Degree/Certification Awarded: M.S.

University of Detroit
Mechanical Engineering Department
4001 W. McNichols
Detroit, MI 48221

Robotics Contact Person: Dr. K. Taraman
Phone: (313) 927-1242
Courses Available: Robotics, Hydraulics, CAD/CAM,
N/C, Microprocessors
Length of Program/Course: M.E. - 1 1/2 yrs.
D.E. - 3 yrs.
Enrollment Requirements: Undergraduate Degree w/
manufacturing experience
Robot Laboratory Available: No
Degree/Certification Awarded: B.M.E., M.E., D.E.

Oakland University
School of Engineering
Rochester, MI 48063

Robotics Contact Person: Prof. Donald Falkenburg
Phone: (313) 377-2218 or 2210
Courses Available: Robotics, CAD/CAM, Integrated
Systems, Microprocessors, Pattern
Recognition, Machine Vision, Artificial Intelligence, Control
Systems
Length of Program/Course: 4 years, 14 weeks
Enrollment Requirements: Undergraduate Degree
Robot Laboratory Available: Yes
Major Laboratory Equipment: PUMA 600, 6 Mini
Movers, 2 Bendix PAC
robots. Labs also in
CAD, Machine Vision,
Pattern Recognition
Degree/Certification Awarded: M.S., Ph.D.

MISSOURI

University of Missouri-Rolla
Mechanical Engineering Dept.
Rolla, MO 65401

Robotics Contact Person: R.T. Johnson
Phone: (314) 341-4614
Program Title: Mechanical Engineering
Courses Available: Robotics, CAD/CAM,
Microprocessors, Logics
Robot Laboratory Available: Yes
Major Laboratory Equipment: PUMA 600, Data
General Nova 3

NEW JERSEY

Rutgers University
P.O. Box 909
Piscataway, NJ 08854

Robotics Contact Person: Dr. Ting W. Lee
Phone: (201) 932-3680
Courses Available: Robotics, CAD/CAM,
Microprocessors
Enrollment Requirements: Undergraduate Degree
Robot Laboratory Available: Under Development
Major Laboratory Equipment: RHINO XR-1 Robot
and control system,
Apple computers
Degree/Certification Awarded: M.S., Ph.D.

NEW YORK

Columbia University
Mechanical Engineering
234 S.W. Mudd
New York, NY 10027

Robotics Contact Person: George Klein
Phone: (212) 280-2955
Courses Available: Robotics, Robot Vision,
Microprocessors

Enrollment Requirements: Acceptance to Program
Robot Laboratory Available: Under Development

NORTH CAROLINA

North Carolina State University
Electrical Engineering
Raleigh, NC 27650

Robotics Contact Person: Wesley Snyder
Phone: (919) 737-2336, Ext. 33
Courses Available: Robotics, CAD/CAM,
Microprocessors
Enrollment Requirements: Acceptance to Program
Robot Laboratory Available: Under Development
Major Laboratory Equipment: Complete robot and robot vision research lab.
Teach lab under
development.

OHIO

Ohio University
Industrial Technology
69 W. Union Street
Athens, OH 45701

Robotics Contact Person: W. H. Creighton
Phone: (614) 594-5300
Program Title: Industrial Technology
Courses Available: Robotics, Hydraulics, Pneumatics,
            CAD/CAM, N/C
Enrollment Requirements: Undergraduate Degree
Robot Laboratory Available: Yes
Major Laboratory Equipment: 1 Robot, 1 CNC
                system, N/C machines

SOUTH CAROLINA

Clemson University
Mechanical Engineering
Riggs Hall
Clemson, SC 29631

Robotics Contact Person: Dr. Frank W. Paul
Phone: (803) 656-3291
Courses Available: Hydraulics/Pneumatics, N/C,
            Microprocessors, Logics
Enrollment Requirements: Acceptance to Program
Robot Laboratory Available: Yes
Major Laboratory Equipment: Versatran E, Microbot,
            Dec 1103, Apple II

TENNESSEE

Vanderbilt University
Electrical & Biomedical Engineering
Nashville, TN 37235

Robotics Contact Person: Gerald Cook
Phone: (615) 322-2771
Program Title: Electrical & Biomedical Engineering
Courses Available: Robotics, Microprocessors
Enrollment Requirements: Undergraduate Degree
Robot Laboratory Available: Under Development

TEXAS

University of Texas-Austin
Mechanical Engineering
Austin, TX 78703

Robotics Contact Person: N. Duke Perreira
Phone: (512) 471-1331
Courses Available: Robotics, CAD/CAM, Integrated
            Systems, Microprocessors, Logics
Length of Program/Course: M.S.=1-2 yrs.
                Ph.D.=3-4 yrs.
Enrollment Requirements: Undergraduate Degree
Robot Laboratory Available: Under Development
Degree/Certification Awarded: M.S., Ph.D.

Texas A & M University
Industrial Engineering
College Station, TX 77843

Robotics Contact Person: Robert E. Young
Phone: (713) 845-5440
Program Title: Industrial Engineering, Mechanical
            Engineering
Courses Available: Robotics, N/C, Integrated Systems
Length of Program/Course: M.S.=36 hrs., Ph.D.=3
                yrs.
Enrollment Requirements: Undergraduate Degree
Robot Laboratory Available: Yes
Major Laboratory Equipment: PUMA 600, FANUC
                robot, TI 990/12
Degree/Certification Awarded: M.S.I.E., M.E.I.E.,
                Ph.D.

WISCONSIN

University of Wisconsin-Madison
Mechanical Engineering
1513 University Avenue
Madison, WI 53706

Robotics Contact Person: K.F. Eman
Phone: (608) 262-5907
Courses Available: Robotics, Hydraulics, CAD/CAM,
            N/C
Enrollment Requirements: Senior in Graduate
                Program
Robot Laboratory Equipment: Under Development
Major Laboratory Equipment: Unimate 1000, 1 CNC
                lathe

CANADA

University of Toronto
Mechanical Engineering
5 Kings College Road
Toronto, Ontario M5S 1A4
Canada

Robotics Contact Person: Dr. A. Goldberg/Dr. R.
                Fenton
Phone: (416) 978-7198
Program Title: Mechanical Engineering
Courses Available: Robotics, Hydraulics, CAD/CAM,
                Microprocessors
Length of Program/Course: 4 years, 14 weeks
Enrollment Requirements: Acceptance to Program
Robot Laboratory Available: Under Development
Major Laboratory Equipment: 1 PUMA

Appendix B

Robotics Timetable

1750	Swiss craftsmen create automatons with clockwork mechanisms to play tunes and write letters.
1817	Mary Shelley writes the novel *Frankenstein*, which depicts a humanoid robot created from parts of human cadavers.
1917	The word *robot* first appears in literature, coined in the play *Opilek* by playwright Karel Capek, who derived it from the Czech word *robotnik*, meaning "slave."
1921	The term *robot* is made famous by Capek's play *R.U.R.* (Rossum's Universal Robots).
1927	German Director Fritz Lang creates the first female robot, depicted as a liberator of factory workers, in his film, *Metropolis*.
1938	Isaac Asimov coins the term *robotics* in his science fiction novels, and formulates the Three Laws of Robotics to prevent robots from harming humans.
1942	Isaac Asimov invents the occupation of *robopsychologist* for Susan Calvin, a character in one of his science fiction novels.
1954	The first United Kingdom robotics patent, No. 781465, is granted in England on March 29.

1956	Robot inventor Joseph Engelberger and George C. Devol meet at a cocktail party to discuss Isaac Asimov's writings about robots, and to discuss building robots of their own.
1956	The Logic Theorist, an artificial intelligence machine capable of proving logical propositions point-by-point, is unveiled at Dartmouth College.
1960	Artificial intelligence teams at Stanford Research Institute (California) and the University of Edinburgh in Scotland begin work on machine vision.
1961	George C. Devol, the "Father of Robotics," obtains the first U.S. robot patent, No. 2,998,237.
1964	The first Trallfa robot is used to paint wheelbarrows in a Norwegian factory during a human labor shortage.
1966	The first prototype painting robots are installed in factories in Byrne, Norway.
1968	Unimation takes its first multi-robot order from General Motors.
1969	Unimate robots assemble Chevrolet Vega automobile bodies for General Motors.
1970	General Motors becomes the first company to use machine vision in an industrial application. The Consight system is installed at a foundry in St. Catherines, Ontario, Canada.
1970	The first American symposium on robots meets in Chicago.
1971	Formation of the Japan Industrial Robot Association (JIRA) makes Japan the first nation to establish a national robotics organization.
1972	The SIRCH machine, capable of recognizing and orienting randomly presented two-dimensional parts, is developed at the University of Nottingham, England.
1972	Kawasaki installs a robot assembly line at Nissan Motors (manufacturer of Datsun automobiles) in Japan.
1973	*The Industrial Robot,* the first international journal of robotics, begins publication.
1973	ASEA introduces its all-electric-drive IRb 6 and IRb 60 robots, designed for automatic grinding operations.

1974	Hitachi uses touch and force sensing with its Hi-T-Hand robot, allowing the robot hand to guide pins into holes.
1975	Cincinnati Milacron introduces its first T3 robot for drilling applications.
1976	The ASEA 60kg robot is the first robot installed in an iron foundry; the Cincinnati Milacron T3 becomes the first robot to enter the aerospace industry.
1976	The Trallfa spray-painting robot is adapted for arc welding at the British agricultural implement firm of Ransome, Sims and Jefferies.
1976	Remote Center Compliance evolves from research at the Charles Stark Draper Laboratories in Cambridge, Massachusetts. Dynamics of part mating are developed, allowing robots to line up parts with holes both laterally and rotationally.
1978	The first PUMA (Programmable Universal Assembly) robot is developed by Unimation for General Motors.
1979	Japan introduces the SCARA (Selective Compliance Assembly Robot Arm); Digital Electronic Automation (DEA) of Turin, Italy, introduces the PRAGMA robot, which is licensed to General Motors.
1980	The biggest changes in robotics occur in control and software. Robotics languages are developed to ease programming bottlenecks.
1981	IBM enters the robotics field with its 7535 and 7565 Manufacturing Systems.
1981-1984	Rehabilitation robots are enhanced by mobility, voice communication, and safety factors. Greater emphasis is placed on machine vision, tactile sensors, and languages. Battlefield and security robots are developed.
1985 and beyond	New computer architectures, artificial intelligence, and natural language programming will be combined in ever more innovative and useful ways. The development of computer-aided design (CAD), computer-aided manufacturing (CAM), and computer-integrated manufacturing (CIM) will proceed at an ever-accelerating rate.

Appendix C

Manufacturer Listings

ROBOT MANUFACTURERS

Acco Industries, Inc.
Babcock International
101 Oakview Drive
Trumbull, CT 06611

Acrobe Positioning Systems, Inc.
3219 Doolittle Drive
Northbrook IL 60062

Admiral Equipment Co.
305 W. North St.
Akron, OH 44303

Advanced Robotics Corp.
Newark Industrial Park
Building 8, Route 79
Hebron, OH 43025

Air Technical Industries
7501 Clover Ave.
Mentor, OH 44060

American Robot Corp.
354 Hookstown Rd.
Clinton, PA 15026

Androbot, Inc.
101 E. Daggett Drive
San Jose, CA 95134

ASEA, Inc.
4 New King St.
White Plains, NY 10604

Automated Assemblies Corp.
Subsidiary of Nypro, Inc.
Clinton, MA 01510

Automation Corp.
Marathon Industries
23996 Freeway Park Drive
Farmington Hills, MI 48024

Automatix, Inc.
1000 Technology Park Drive
Billerica, MA 01821

Bendix Corp.
Robotics Division
21238 Bridge St.
Southfield, MI 48034

Binks Mfg. Co.
9201 W. Belmont Ave.
Franklin Park, IL 60131

Bra-Con Industries, Inc.
12001 Globe Rd.
Livonia, MI 48150

Cincinnati Milacron, Inc.
Industrial Robot Division
215 S. West St.
Lebanon, OH 45036

Comet Welding Systems
900 Nicholas Blvd.
Elk Grove Village, IL 60007

Control Automation, Inc.
P.O. Box 2304
Princeton, NJ 08540

Cybotech, Inc.
Division of Ransburg Corp.
P.O. Box 88514
Indianapolis, IN 46208

Cyclomatic Industries, Inc.
Robotics Division
7520 Convoy Ct.
San Diego, CA 92111

Defense Advanced Research Projects Agency
(DARPA)
1400 Wilson Blvd.
Arlington, VA 22209

Denning Mobile Robots
22 Cummings Park
Woburn, MA 01801

DeVilbiss Co.
Div. of Champion Spark Plug Co.
300 Phillips Ave.
P.O. Box 913
Toledo, OH 43692

ESAB/Heath
P.O. Box 2286
Ft. Collins, CO 80522

Expert Automation, Inc.
(KUKA)
40675 Mound Rd.
Sterling Heights, MI 48078

Gallaher Enterprises
2110 Cloverdale Ave., Suite 2B
Winston-Salem, NC 27103

GCA CORP.
PaR SYSTEMS
3460 Lexington Ave. No.
St. Paul, MN 55112

General Electric Co.
Automation Systems Operation
1285 Boston Ave.
Bridgeport, CT 06602

General Numeric Corp.
390 Kent Ave.
Elk Grove Village, IL 60007

Graco Robotics Inc.
Graco, Inc.
12898 Westmore Ave.
Livonia, MI 48150

Hall Automation
c/o Fared Robot Systems, Inc.
3860 Revere St., Suite D
Denver, CO 80239

Hirata Corp. of America
8900 Keystone Crossing
Indianapolis, IN 46240

Hitachi America Ltd.
6 Pearl Ct.
Allendale, NJ 07401

Hobart Brothers Co.
600 W. Main St.
Troy, OH 45373

Hodges Robotics International Corp.
3710 N. Grand River Ave.
Lansing, MI 48906

Hubotics, Inc.
6352 Corte del Abeto
Carlsbad, CA 92008

Intelledex, Inc.
33840 East Gate Circle
Corvallis, OR 97333

International Business Machines Corp.
Advanced Manufacturing Systems
1000 N.W. 51st St.
Boca Raton, FL 33432

International Robomation/Intelligence
2281 Las Palmas
Carlsbad, CA 92008

C. Itoh & Co. America, Inc.
21415 Civic Center Drive
Southfield, MI 48076

Jet Propulsion Laboratory
California Institute of Technology
4800 Oak Grove Drive
Pasadena, CA 91103

Lloyd Tool & Manufacturing Corp.
2475 E. Judd Rd.
Burton, MI 48529

Machine Intelligence Corp.
330 Potrero Ave.
Sunnyvale, CA 94086

Manca Inc.
Leitz Bldg.
Link Drive
Rockleigh, NJ 07647

Martin Industries, Div.
Westovver Air Park
Chicopee, MA 01022

Martin-Marietta Corp.
Robotics Division
Middle River
Baltimore, MD 21220

Mentor Products, Inc.
7763 Mentor Ave.
Mentor, OH 44060

Microbot, Inc.
453-H Ravendale Drive
Mountain View, CA 94043

Midway Machine & Engineering Co.
2324 University Ave.
St. Paul, MN 55114

Mobot Corp.
980 Buenos Ave.
San Diego, CA 92110

Nordson Corp.
Robotics Division
555 Jackson St.
P.O. Box 151
Amherst, OH 44001

Pedsco
12 Principal Road, Unit 2
Scarborough, Ontario, Canada M1R 4Z3

Pentel of America
1100 Arthur Ave.
Elk Grove Village, IL 60007

Perceptronics
6271 Variel Ave.
Woodland Hills, CA 91367

PickOmatic Systems, Inc.
Div. of Fraser Automation
37950 Commerce Drive
Sterling Heights, MI 48077

Planet Corp.
Robot Division
27888 Orchard Lake Rd.
Farmington Hills, MI 48018

Prab Robots, Inc.
59144 E. Kilgore Rd.
Kalamazoo, MI 49003

R2000 Corp.
804 Broadway
West Long Branch, NJ 07764

Reis Machines
1426 Davis Rd.
Elgin, IL 60120

Rhino Robots, Inc.
P.O. Box 4010
Champaign, IL 61820

Rimrock Corporation
1700 Rimrock Road
Columbus, OH 43219

Rob-Con Ltd.
12001 Globe Rd.
Livonia, MI 48150

Robot Defense Systems, Inc.
3860 Revere St.
Denver, CO 80239

Robot Systems, Inc.
Corporate Division
50 Technology Parkway
Technology Park Atlanta,
Norcross, GA 30092

The Robot Factory
P.O. Box 112
Cascade, CO 80809

Robotics, Inc.
RD 3, Rte. 9
Ballston Spa, NY 12020

Robotics International Corp.
2335 East High St.
Jackson, MI 49203

Robo-Vend, Inc.
2138 Woodson Rd.
St. Louis, MO 63114

Sandhu Machine Design, Inc.
308 State St.
Champaign, IL 61820

Seiko Instruments USA, Inc.
2990 W. Lomita Blvd.
Torrance, CA 90506

Scovill Manufacturing Co.
Schrader Bellows Div.
200 W. Exchange St.
Akron, OH 44309

Snow Mfg/Origa Corp.
928 Oaklawn
Elmhurst, IL 60126

Southwest Research Co.
6220 Culebra St.
San Antonio, TX 78238

Sterling Detroit Co.
261 E. Goldengate Ave.
Detroit, MI 48203

Textron, Inc.
Bridgeport Machines Div.
500 Lindley St.
Bridgeport, CT 06606

Thermwood Corp.
P.O. Box 436
Dale, IN 47523

Tokico America, Inc.
3555 Lomita Blvd.
Torrance, CA 90505

Unimation, Inc.
Shelter Rock Lane
Danbury, CT 06810

United States Robots, Inc.
1000 Conshohocken Rd.
Conshohocken, PA 19428

United Technologies Corp.
Steelweld Robotics Div.
5200 Auto Club Drive
Dearborn, MI 48126

Westinghouse Electric Corp.
Industry Automation Div.
400 Media Drive
Pittsburgh, PA 15205

Yaskawa Corp. of America
305 Era Drive
Northbrook, IL 60062

MACHINE VISION SYSTEMS MANUFACTURERS

Analog Devices
Norwood, MA 02062

ASEA, Inc.
White Plains, NY 10604

Automated Vision Systems
Campbell, CA 95008

Automatix, Inc.
Burlington, MA 01803

Control Automation, Inc.
Princeton, NJ 08540

Diffracto Ltd.
Windsor, Ontario, Canada

General Electric Co.
Optoelectronic Systems Operation
Syracuse, NY 13201

International Imaging Systems
Milpitas, CA 95035

Machine Intelligence Corp.
Sunnyvale, CA 94088

Object Recognition Systems, Inc.
Princeton, NJ 08540

Octek, Inc.
Burlington, MA 01803

Perception, Inc.
Farmington Hills, MI 48018

Robotic Vision Systems, Inc.
Melville, NY 11747

Selecom, Inc.
Valdese, NC 28690

Vuebotics Corporation
Carlsbad, CA 92008

Further information on robotics and computer-integrated manufacturing may be obtained from:

Computer-Aided Manufacturing International
611 Ryan Plaza Drive
Suite 1107
Arlington, TX 76011

Parts of this list of manufacturers of robots and machine vision systems were obtained from *Robotics Sourcebook and Dictionary,* by David F. Tver and Roger W. Bolz. Industrial Press, Inc. New York, 1983, pp. 135-138.

Robotics Glossary

access time—An interval of time between the moment when data are requested from a computer and the instant when delivery arrives.

accuracy—Quality of conformance to a standard. *Absolute accuracy* is the difference between a point indicated by a robot's control system and the point achieved by the manipulator. *Repeat accuracy* is the cycle-to-cycle variation of the manipulator arm as it is aimed at the same point. *Measurable accuracy* is the degree to which the actual position corresponds to the desired position.

acoustic coupler—An electronic machine that sends and receives digital data through a standard telephone handset.

active accommodation—Sensors, control, and robot motion are integrated to alter a robot's preprogrammed motions in response to sensed forces. Active accommodation stops or produces feedback from a robot when forces reach certain levels.

active illumination—A controlling source emits light of varying intensity, direction, and projection.

actuator—The actuator delivers power for robot motion. It consists of an electrical, hydraulic, or pneumatic motor, transducer, or cylinder that acts as a driver.

adaptable—A robot's adaptability is determined by its vision system, force, and tactile sensors which enable it to make self-directed corrections with little human intervention.

adaptive control—A method by which input from sensors changes in an attempt to achieve better performance. Control parameters are automatically adjusted.

ADCCP (Advanced Data Communications Control Procedure)—The proposed American national standard for computerized data communications protocol. The procedure utilizes a cyclic redundancy error check method.

133

analog control—A robot may be directed by electric, hydraulic, pneumatic, and other analog signal processing devices used as control systems.

analog-to-digital (A/D) converter—An electronic device which takes a voltage signal (also called an *analog* signal) and transforms it into a digital signal (a string of binary ones and zeros) for use by a digital computer.

analogic control—A robot can be controlled by communication signals which are isomorphic (correspond one-to-one) to the variables being controlled by a human operator. An analogic control device performs such a role.

android—A robot that looks like a human.

annunciator—A light or sound signal on a robot that flashes or makes noise to attract attention and alert humans that the robot is in operation.

architecture—The basic design and structure of a robot, computer, or manufacturing process.

arm—The joints, links, slides, and mechanical design of a robot which support a moving tool attachment or hand.

artificial intelligence—The ability of a robot or computer to perform tasks and functions normally attributed only to humans. Some examples are decision making, perception, problem solving, pattern recognition, cognition, understanding, teaching, and learning.

ASCII—American Standard Code for Information Interchange, a standardized coding method for representing alphanumeric characters and other symbols as 8-bit (one-byte) computer "words."

assembly language—A computer language used to control some robots, consisting of short, symbolic expressions that represent in human-readable form, the binary-level machine language used by a computer or processor.

assembly robot—A robot designed, programmed, or dedicated to putting together parts into complete products or subassemblies. These robots are small, fast machine tools used for grasping and fitting parts together.

automation—The method of making a process self-controlling.

azimuth—The angular distance from an observer's line of view to a point in a horizontal plane.

background processing—The processing of low-priority computer programs only in time not used by higher-priority programs.

band—A group of tracks on a magnetic drum or disk called the *frequency spectrum.*

bang-bang control—A control system which quickly changes from one state to another.

bang-bang robot—A non-servo-controlled robot that bangs into fixed stops. Each axis is driven against a mechanical stop.

base—The fixed platform to which a robot's shoulder is attached.

base address—The numeric value used as a reference point when calculating addresses within a computer program.

base number—The number which is the basis for counting in any number system, sometimes called the *radix* of the system.

BASIC—Beginners All Purpose Symbolic Instruction Code, a computer programming language used to program some robots and computers.

batch manufacturing—The process by which a factory produces groups of parts in which each part is identical.

Bel character—The ASCII character that controls the program when there is a need to ask for human intervention and that may activate other attention-getting devices or alarm systems.

bilateral manipulator—A master-slave manipulator (two-armed) with symmetric force reflection, where both master and slave arms have sensors and actuators. The arms use equal and opposing forces and coordinated movements to perform a task.

binary—Having only two parts, used specifically of a number system in which only the digits one and zero are used to represent numbers.

binary arithmetic operation—Representation of the operands and the result of an arithmetic operation in pure binary numeration.

binary code—The combination of two distinct characters (0 and 1, in most cases) used to create computer programs or languages.

binary coded decimal (BCD)—A system of information coding in which individual decimal digits are represented by a pattern of ones and zeros. In addition, any decimal number from 0 to 9 can be represented by a combination of four values of 1,2,4, and 8.

binary picture—A black-and-white video screen image that shows no shades of grey.

bit—In a digital computer system, a bit (short for *binary digit*) is the smallest piece of data with which the computer can operate.

branching—The process of changing the normal sequential execution of statements in a program.

breakaway force—The resistive force, such as static friction, that is not constant as the relative velocity increases.

bus—A common data channel shared by several digital signals, transmitted using an information coding scheme by which the different signals are identified. Also, a common power distribution system that can be used as a source by several devices.

byte—A unit of computer-stored information or the name of a string of binary numbers, usually 8 bits long.

cable drive—A robotic system wherein power is transmitted from an actuator to a remote mechanism by the use of cables and pulleys.

CAD (computer-aided design)—The use of a computer to design or change a product, such as using a computer for drafting or illustrating.

calibration—The act of determining the scale graduations of an instrument/machine or rectifying the deviations from a standard in order to ascertain the best corrections.

call—The method of bringing a computer routine into effect by keyboarding the entry conditions or invoking a program command.

CAM (computer-aided manufacturing)—The use of computer technology to direct and control the manufacturing process.

cartesian coordinate system—A coordinate system originating from an intersection of three perpendicular straight lines (axes).

card code—The mixture of punched holes used to represent data on a punch card (sometimes called a *Hollerith card*) used for computer data input.

cathode ray tube (CRT)—A device that depicts data visually by means of a controlled electron beam hitting a phosphorescent surface. Among other places, it is found behind the video screens of computer monitors and television sets.

CCD camera—A solid-state television camera which uses charge-coupled device (CCD) technology.

cell—Two or more work stations within a factory that share the interconnecting material transport mechanisms and storage buffers.

cell control—A module in the ICAM control hierarchy that directs a cell.

center—An ICAM manufacturing unit consisting of a number of cells and the materials transport/storage that interconnect the two.

center of gravity—A point in a body at which the mass of the body can be concentrated to produce the same gravity as the entire body.

centralized control—Control decisions for two or more control jobs at various locations made at a centralized location.

central processing unit (CPU)—The "brain" of a computer which executes instructions and operates by the inflow of data programs.

chain drive—Transmission of power from an actuator to a remote mechanism by means of mating tooth sprocket wheels with an adaptable chain.

channel—A path over which information is transmitted.

chip—The semiconductor (silicon) material on which integrated circuits are imprinted.

CID camera—A solid-state camera that uses a charge-injection imaging device (CID) to convert an image into digital form in order to be processed by a computer.

CIM (computer-integrated manufacturing)—The use of computers and robots integrated for total factory automation.

circuit—A communication link between several points.

closed loop—A loop that has no exit and whose execution can be interrupted only by intervention from outside the computer.

closed-loop control—Control achieved by measuring the degree to which actual system response conforms to a preferred response feedback. The difference between the system response and the preferred response can be used to direct the system to conform to the correction.

closed subroutine—A subroutine that can be stored in one area and linked to one or more calling routines.

CMOS (complementary metal-oxide semiconductor)—A type of integrated circuit which combines thousands of components on a single chip and is characterized by low power dissipation and moderate circuit speed.

CNC (computer numerical control)—The use of a computer to direct mechanical equipment.

code—A rule or regulation used to change data from one representation to another. Codes express data in numbers or letters in order to reduce storage space in a computer program.

communications link—A path over which information can be sent from one point to another, usually in the form of a wire or optical fiber.

compiler—A program that changes a high-level computer language into a machine language composed of binary code (ones and zeros) so that it can be read by a computer.

computed path control—A control design that allows the path of the manipulator end point to be computed so that the result achieved conforms to a desired criterion such as a minimum time or other limit.

computer graphics—The input of pictorial data such as off-line drawings and photographs from scanners, digitizers, and pattern-recognition devices, and the output of drawings on paper, film, and video screens.

computer numerical control—Employing a computer using floppy disk or cassette tape to implement the numerical control function of a device, machine, or other computer program.

conditional statement—A computer program direction whose execution depends upon other criteria being fulfilled.

coordinated axis control—The end point control achieved when the axes of a robot arrive at their end points at the same time, giving a smooth appearance to the motion as the end point moves along a programmed path.

CPU (central processing unit)—The part of a computer that executes instructions as data is received.

cycle time—A pattern of continuous repetition beginning when one machine operation starts and ending when another machine operation begins.

cylindrical coordinate robot—A robot whose degrees of freedom of its manipulator arm are defined in cylindrical coordinates.

data bank—A centralized computer storage facility which contains information on a subject, usually referred to as the collection of libraries used by a business.

database—A collection of information comprised of comprehensive files in one or more areas of similar interest.

degree of disorder—The amount of divergence from an orderly environment where parts must always be in the same attitude, orientation, and position.

degree of freedom—A way in which a body or point may move or change by an independent variable.

diagnostic—The troubleshooting, detection, isolation, and callibration of an error or malfunctioning machine, part, component, or instrument.

drift—The tendency of a robot to move away from a programmed response.

duty cycle—The time interval during which a machine or system will be at full power.

echo check—The technique of verifying the accuracy of transmitted data by returning the received data to the sending end for comparison with the original data.

editor program—A computer program designed to modify and delete data in accordance with prescribed rules.

encoder—A transducer used to convert angular position to digital data.

end effector—A mechanical device, attached to the wrist of a manipulator, with which objects can be grasped or gripped

end-of-arm speed—Depending on its position, the load carried, and the axes about which the robot arm moves, the end-of-arm speed can be computed by asking how fast the gripper can go from one point to another in the envelope. The result is usually a ballpark figure based upon a differential equation.

envelope—The boundary of a work station area; the boundary of a robot's reach within a work cell or work station.

error control procedure—The method of checking sequence numbering to detect errors, and/or detecting and recovering from errors occurring in transmitted data.

exoskeleton—An articulated mechanism whose joints are similar to those of a human arm and that will move in correspondence to a human operator's hand when attached to a human arm.

fail safe—Failure of a robot or system without danger or damage to humans and products.

fail soft—Failure of a robot or system without interruption of performance or product quality.

fembot—A humanoid robot given female sex characteristics or a female orientation.

file—A bank or library of organized records of information.

fixed-stop robot—A robot having a fixed limit at each end of its stroke on each of its axes, and which cannot stop except at one or another of these limits.

gripper—A robot hand that is capable of grasping, holding, and manipulating parts, components, and other objects.

group technology—A system for coding parts based on similar characteristics.

hand—A metal device attached to a robot's arm that contains a means to grasp objects.

hexadecimal—A number system with a base of 16, using the digits 0, 1, 2, 3, 4, 5, 6, 7, 8, 9, A, B, C, D, E, and F.

hierarchy—A relationship of parts giving those at higher levels priority over those at lower levels.

hydraulic motor—An actuator having interconnected parts which convert high-pressure hydraulic or pneumatic fluid into mechanical rotation or shaft movement.

induction motor—An alternating current motor that produces torque when a rotating magnetic field meets electric current induced in circuits or coils.

industrial robot—A programmable, multifunction manipulator built to transport parts/tools through programmed steps.

integrated circuit—An electronic circuit constructed on a semiconductor chip.

intelligent robot—A programmable machine tool that can make performance choices based on sensory inputs.

interface—A connection whereby two people or objects share a boundary. This can refer to two computers being interconnected, an operator programming a computer or robot, or two people sharing information.

job shop—A manufacturer who mixes products in low volume production of batch lots.

joint—A robot's rotational or other degree of freedom, or the degree of freedom in any manipulator or system.

joystick—A flexible handle used by a human operator to operate a robot, computer, or other machine.

large-scale integration (LSI)—A classification of the complexity of an integrated electronic circuit chip. LSI usually is considered to be the equivalent of 100 or more transistors on a single chip.

learning control—A control scheme whereby experience is used to create future control decisions which will surpass those used before.

level of automation—The measurable level of mechanization of a process; the degree to which a process can be done by machines as opposed to humans.

light-emitting diode (LED)—A semiconductor device that emits light when electricity passes through it.

limited-degree-of-freedom robot—A robot able to position and orient itself in fewer than six degrees of freedom.

limit switch—An electrical switch that flips when a moving part reaches its programmed limit, thereby stopping motion.

linearity—The degree to which the graph of a function or process resembles a straight line.

load—An external force applied to a body.

load capacity—The greatest weight or dimensions of an object that can be handled by a machine or process without failure of the equipment.

load deflection—The difference in position of a tool between a non-loaded and externally loaded condition.

loop—A sequence of directions that repeats until some condition is met.

machine language—A computer language usually specific to and used directly by a particular machine without further programming in a higher-level language.

macro—A sequence of instructions generated and defined under a single name.

manipulation—The ability of a robot to grasp and lift an object.

manipulator—A remotely controlled or human-operated device consisting of a series of jointed or sliding segments used to grasp and move objects.

master/slave manipulator—A teleoperator having geometrically isomorphic arms. The master is positioned by a human; the mechanical slave arm duplicates the human's movements.

microprocessor—The brains or main processing part of a microcomputer built as a single integrated circuit.

mobile robot—A robot that is either mounted on a movable platform or that has walking or vehicular ability.

modem—A data-transmitting device connecting a computer to telephone lines in order to send and receive digital data. The word *modem* is short for *modulator-demodulator*.

monitor—A computer peripheral device that observes and controls or checks the operations of a computer system or robot.

net load capacity—The weight and mass of an object that can be handled by a machine or process without failure, over and above that required for a device that accompanies the object.

numerical control—A technique that automatically controls a machine tool by using prerecorded information.

open loop control—Direction achieved by driving actuators with a sequence of preprogrammed signals without measuring system response and closing the feedback loop.

open-loop robot—A robot that has no means of comparing actual output feedback to input regarding the robot's position or rate.

operating system—The software which controls and directs a computer or robot is referred to as a computer's *operating system*.

part classification—Products are coded according to part family by four or more digits in a coding scheme.

pattern recognition—The classification of data or pictures into set categories used in artificial intelligence; the ability of a robot or computer to see objects or characteristics within a picture; the ability of a machine to recognize shapes and distinguish objects from shadows.

pick-and-place robot—A robot with less than three degrees of freedom which transfers objects by means of point-to-point moves; a bang-bang robot.

pixel—A light-sensitive point or array of points on a computer video screen that can be turned on or off by the scanning electron beam to form an entire picture.

point-to-point control—A scheme whereby the robot's programming commands specify a limited number of points along a path.

proximity sensor—A device placed on a robot to sense whether an object is nearby, preventing it from smashing into the object. Proximity sensors utilize light, sound, eddy currents, magnetic fields, air jet pressure, and other methods of detecting objects in the robot's path.

rate control—Control system in which the preferred input is the velocity of the controlled object.

record-playback robot—A manipulator that plays back trajectory points to the robot servo system. The critical points are stored in sequence by recording the actual values of the robot's joint position encoders.

resolution—The smallest interval between two adjacent details that can be distinguished from one another.

resolved motion rate control—A control scheme whereby a robot's axes are coordinated in order that the velocity of the end point is under direct control.

robot—A programmable mechanical device that moves under automatic control.

robot programming language—A computer language designed for the creation of programs to be used for controlling robots.

robotnik—Original Czech term for robot, meaning "slave."

routing—The sequence of operations in production executed to build a part or an assembly. In telecommunications, *routing* refers to the communication path taken by a message to reach its destination.

SCR (Silicon Controlled Rectifier)—A solid-state electronic control mechanism by which small electrical currents can be used to control larger electrical loads.

SDLC (Synchronous Data Link Control)—A bit-oriented method for managing the flow of information within a data communications link.

semiconductor—A solid material such as germanium or silicon material in which the transmission of electricity can be controlled by external signals.

sensorycontrolled robot—A robot whose motions are controlled by information sensed from the place in which it operates.

sequence robot—A robot whose physical motion and path follow a prerecorded sequence.

servo-controlled—Directed by a driving signal which is determined by the error between the machine's present position and the desired output position.

servo mechanism—A mechanically or electrically controlled device whose driving signal is determined by the difference between the commanded position and the realistic position.

servo valve—A valve which allows a flow of a hydraulic fluid proportional to a low-level signal input to the valve.

shoulder—The joint of an industrial robot that comes between the arm and the base.

smart sensor—A device in which the output signal depends on logical operations that combine sensory information with input from areas outside the sensor.

software—The instructions, programs, and formulas which are used to make computers operate in logical sequence.

solid-state camera—A television camera that utilizes a solid-state integrated circuit to convert light into electronic signals.

spherical coordinate robot—A robot that is built with a horizontally rotating base, a vertically rotating shoulder, and a linear traversing arm to produce a circular motion.

spot welding—A method of putting together sheet metal parts using electric current. The electricity is transmitted via two electrodes to a point where metal sheets are to be joined, melting the sheets together to form a circular spot weld.

stepping motor—An electric motor that rotates in discrete steps rather than continuous movement.

tachometer—A mechanism that senses the speed at which a shaft rotates.

tactile sensor—A device on the hand or gripper of a robot which senses physical contact with an object. A tactile sensor prevents the robot from using too much power to crush the object. It gives a robot some sense of touch in relation to an object.

target language—The language into which statements are translated.

teleoperator—A master/slave device which produces movements in a robot by remote control from a human operator. The teleoperator gives feedback to the human operator. It is used extensively with military, security, mobile, and industrial robots.

transducer—A device which converts one form of electrical, light, or mechanical energy into another.

transfer machine—A machine designed to grasp a work piece and move it from one stage of a manufacturing operation to another.

transformation—A mathematical conversion system used in robotics to give industrial robots line tracking ability.

TTL (Transistor-Transistor Logic)—A signal processing method allows data to be processed through circuits in the form of low-level electric signals.

undershoot—The degree to which a system's response to a step change falls short of the desired value.

upper arm—The portion of a robot's jointed arm which is connected to the shoulder.

vidicon—One type of electron tube used in television cameras to change an optical image into an electrical signal.

volatile memory—A memory system in a control system which requires a continual source of electric current to keep the data in storage.

work station—A manufacturing unit consisting of numerically controlled machine tools serviced by a robot.

working envelope—The maximum reach of a robot's hand or working tool in all directions, represented by a set of points.

working range—All positions inside the working envelope.

working space—The space bounded by the working envelope.

wrist—A robot's set of rotary joints between the arm and the hand which allows the gripping action to be oriented in different directions.

yaw—An angular displacement viewed along the principal axis of an object having a top side, especially along its line of motion.

zero-address instruction—An instruction that contains no address part. It is utilized when the address is implicit or no address is required.

zero suppression—The elimination of zeros from a number when such zeros have no significance to the value of the numerical statement.

This glossary was prepared from the following sources:

Ritchie, David. *The Binary Brain.* Boston: Little, Brown & Co., 1984.

Smith, Bradford M.; Sheridan, Thomas B.; Albus, James S.; Barbera, Anthony J.; and Vanderbrug, Gordon J. *A Glossary of Terms for Robotics—Revised.* Washington, D.C.: National Bureau of Standards, and Dearborn, Mich.: Society of Manufacturing Engineers, 1983.

Susnjara, Ken. *A Manager's Guide to Industrial Robots.* Englewood Cliffs, N.J.: Prentice-Hall, 1982.

Tver, David F., and Bolz, Roger W. *Robotics Sourcebook and Dictionary.* New York: Industrial Press, Inc., 1983.

Publications of the United States Department of Commerce and the National Bureau of Standards.

Research Bibliography

This bibliography includes material from a recommended reading list compiled jointly by Robotics International of the Society of Manufacturing Engineers and PRAB Robots, Inc., as well as the "Robot Related Reading List" compiled by the Robotics Special Interest Group of the San Diego (California) Computer Society. Both are reprinted here with the permission of their respective compilers.

BOOKS

Allan, John J. *A Survey of Industrial Robots.* Dallas: Productivity International, Inc., 1980.

Arden, Bruce W., ed. *What Can Be Automated?* Cambridge, Mass.: The MIT Press, 1983.

Ashley, Mike, ed. *Souls in Metal: An Anthology of Robot Futures.* New York: Jove Publications, 1977.

Asimov, Isaac. *I, Robot.* Garden City, N.Y.: Doubleday, n.d.

Brod, Craig. *Technostress: The Human Cost of the Computer Revolution.* Menlo Park, Calif.: Addison-Wesley, 1984.

Brodie, Leo. *Starting FORTH.* Englewood Cliffs, N.J.: Prentice-Hall, 1981.

Chen, Wayne. *The Year of the Robot.* Beaverton, Oreg.: dilithium Press, 1981.

Dacosta, Frank. *How to Build Your Own Robot Pet.* Blue Ridge Summit, Pa.: TAB Books, 1979.

Engelberger, Joseph F. *Robotics in Practice*. New York: American Management Assn., Inc., 1980.

Feigenbaum, Edward A., and McCorduck, Pamela. *The Fifth Generation*. Menlo Park, Calif.: Addison-Wesley, 1983.

Heiserman, David L. *How to Build Your Own Self-Programming Robot*. Blue Ridge Summit, Pa.: TAB Books, 1979.

——————. *Build Your Own Robot*. Blue Ridge Summit, Pa.: TAB Books, 1976.

Holland, John M. *Basic Robotic Concepts*. New York: Howard W. Sams & Co., 1983.

Krutch, John. *Experiments in Artificial Intelligence for Small Computers*. New York: Howard W. Sams & Co., 1981.

Loofbourrow, Tod. *How to Build a Computer-Controlled Robot*. Rochelle Park, N.J.: Hayden Book Co., 1978.

Lundstrom, G., ed. *Industrial Robots, A Survey: Details of Construction, Performance, Prices, and Applications*. Great Neck, N.Y.: Scholium International, 1972.

Malone, Robert. *The Robot Book*. New York: Jove Publications, 1978.

National Aeronautics and Space Administration. *Machine Intelligence and Robotics: Report of the NASA Study Group*. Pasadena, Calif.: Jet Propulsion Laboratory, 1979.

Ritchie, David. *The Binary Brain: Artificial Intelligence in the Age of Electronics*. Boston: Little, Brown & Co., 1984.

Robillard, Mark J. *Microprocessor-Based Robotics*. [*Intelligent Machines Series,* Vol. 1]. New York: Howard W. Sams & Co., 1983.

Safford, Edward L., Jr. *Handbook of Advanced Robotics*. Blue Ridge Summit, Pa.: TAB Books, 1983.

——————. *The Complete Handbook of Robotics.* Blue Ridge Summit, Pa.: TAB Books, 1978.

Society of Manufacturing Engineers. *Robots 6: Conference Proceedings and Supplement*. Dearborn, Mich.: Society of Manufacturing Engineers, 1982.

Susnjara, Ken. *A Manager's Guide to Industrial Robots*. Englewood Cliffs, N.J.: Prentice-Hall, 1982.

Tanner, William. *Industrial Robots*. 2d ed. 2 vols. Dearborn, Mich.: Society of Manufacturing Engineers, 1981.

Tver, David F., and Bolz, Roger W. *Robotics Sourcebook and Dictionary*. New York: Industrial Press, 1983.

Ullrich, Robert A. *The Robotics Primer*. Englewood Cliffs, N.J.: Prentice-Hall, 1983.

Weinstein, Martin B. *Android Design*. Rochelle Park, N.J.: Hayden Book Co., 1981.

TECHNICAL PAPERS

Unless noted, all papers listed below are published by Robotics International of the

Society of Manufacturing Engineers, Dearborn, Michigan.

A Glossary of Terms for Robotics—Revised (1983).

Career Opportunities in Robotics and Educational Requirements (1983).

Factors Affecting Management's Resistance to Installation of a Robotics System (1983).

Human Factors Issues in the Factory Integration of Robotics (1982).

Idiosyncrasies of the United States Industrial Robot Market (1983).

JPL Robotics Fact Sheet. (Pasadena Calif.: Jet Propulsion Laboratory, 1980.)

NASA's Use of Robots in Building the Space Shuttle (1983).

Return on Robots (1982).

Robotics and the Automated Manufacturing Industry (1983).

Robotics in the U.K.—An Overview (1980).

Robotics—The Development of the Second Industrial Revolution (1982).

Robot Implementation and the Domino Effect (1982).

The Unattended Factor: FANUC's New Flexibility-Automated Manufacturing Plant Using Industrial Robots (1983).

Trends in the Robotics Industry (1982).

Trends in the Robotics Industry (Revisited): Where Are We Now? (1983).

MAGAZINE ARTICLES

"Costs Down, 'Smarts' Up in New Material Handling Robot." *Material Handling Engineering,* September 1981, 100-104.

"Industrial Robots Today." *Machine and Tool Blue Book,* March 1980, 58-75.

"Just How Many Robots Are Out There?" *American Machinist,* December 1981, 120.

"Now Will People Work with Robots?" *Material Handling Engineering,* November 1981, 35-36.

"Robotics." *Industrial Engineering,* November 1981, 14.

"Robots Join the Labor Force." *Business Week,* 9 June 1980, 62-76.

"Robots Swing into the Industrial 'Arms' Race." *Iron Age,* 21 July, 1980.

"Robots That Assemble." *American Machinist,* November 1981, 175-190.

"Robots That Can Create Jobs." *American Machinist,* January 1982, 131-134.

"The CEO Befriends the Robot—But Manufacturing Must Make It Work." *Production,* September 1981, 88-97.

"The Road to the Automatic Factory 1970-1981." *Manufacturing Engineering,* January 1982, 209-251.

"The Robot Revolution." *Time,* 8 December 1980, 72-83.

"The Robot's Role in Productivity." *Production Engineering,* December 1981, 18-19.

MAGAZINES, JOURNALS, AND DATABASES

Industrial Robots International
Technical Insights, Inc.
158 Linwood Plaza, Box 1304
Fort Lee, NJ 07024

International Journal of
Robotics Research
MIT Press Journals
26 Carleton Street
Cambridge, MA 02142

RBOT Database
BRS
1200 Route 7
Latham, NY 12110
(518) 783-1161
(800) 833-4707

Robot Builder Newsletter
International Robotics Foundation
P.O. Box 3227
Seal Beach, CA 90740

Robotics Today
One SME Drive, P.O. Box 930
Dearborn, MI 48128

Robot News International
The Industrial Robot
(especially the 1983 tenth anniversary
   "Decade of Robotics" issue)
IFS Publications LTD.
35/39 High Street
Kempston, Bedford MK42 7BT, England

Robot/X News
P.O. Box 450
Mansfield, MA 20248

PARTS SOURCES AND CATALOGS

Colne Robotics
207 North 33rd Street
Fort Lauderdale, FL 33334

Hobby Robot Company
P.O. Box 997
Lilburn, GA 30247

Robot Mart
19 West 34th Street
New York, NY 10001

Index

OTHER POPULAR TAB BOOKS OF INTEREST

The Computer Era —1985 Calendar Robotics and Artificial Intelligence (No. 8031—$6.95 paper)

Making CP/M-80® Work for You (No. 1764—$9.25 paper; $16.95 hard)

Going On-Line with Your Micro (No. 1746—$12.50 paper; $17.95 hard)

Mastering Multiplan™ (No. 1743—$11.50 paper; $16.95 hard)

How to Document Your Software (No. 1724—$13.50 paper; $19.95 hard)

Fundamentals of TI-99/4A Assembly Language (No. 1722—$11.50 paper; $16.95 hard)

Computer Programs for the Kitchen (No. 1707—$13.50 paper; $18.95 hard)

Beginner's Guide to Microprocessors—2nd Edition (No. 1695—$9.25 paper; $14.95 hard)

The First Primer of Microcomputer Telecommunications (No. 1688—$10.25 paper; $14.95 hard)

TI-99/4A Game Programs (No. 1630—$11.50 paper; $17.95 hard)

Using and Programming the TI-99/4A, including Ready-to-Run Programs (No. 1620—$10.25 paper; $16.95 hard)

Forecasting on Your Microcomputer (No. 1607—$15.50 paper; $21.95 hard)

Database Manager in MICROSOFT BASIC (No. 1567—$12.50 paper; $18.95 hard)

Troubleshooting and Repairing Personal Computers (No. 1539—$14.50 paper; $19.95 hard)

25 Graphics Programs in MICROSOFT BASIC (No. 1533—$11.50 paper; $17.95 hard)

The Handbook of Microcomputer Interfacing (No. 1501—$15.50 paper)

From BASIC to Pascal (No. 1466—$11.50 paper; $18.95 hard)

Computer Peripherals That You Can Build (No. 1449—$13.95 paper; $19.95 hard)

Machine and Assembly Language Programming (No. 1389—$10.25 paper; $15.95 hard)

55 Advanced Computer Programs In BASIC (No. 1295—$9.95 paper; $16.95 hard)

Programming with dBASE II® (No. 1776—$16.50 paper; $26.95 hard)

Lotus 1-2-3™ Simplified (No. 1748—$10.25 paper; $16.95 hard)

The Last Word on the TI-99/4A (No. 1745—$11.50 paper; $16.95 hard)

The Master Handbook of High-Level Microcomputer Languages (No. 1733—$15.50 paper; $21.95 hard)

Getting the Most from Your Pocket Computer (No. 1723—$10.25 paper; $14.95 hard)

Scuttle the Computer Pirates: Software Protection Schemes (No. 1718—$15.50 paper; $21.95 hard)

MicroProgrammer's Market 1984 (No. 1700—$13.50 paper; $18.95 hard)

PayCalc: How to Create Customized Payroll Spreadsheets (No. 1694—$15.50 paper; $19.95 hard)

How To Create Your Own Computer Bulletin Board (No. 1633—$12.50 paper; $19.95 hard)

Does Your Small Business Need A Computer? (No. 1624—$18.95 hard)

Microcomputers for Lawyers (No. 1614—$14.50 paper; $19.95 hard)

BASIC Computer Simulation (No. 1585—$15.50 paper; $21.95 hard)

Solving Math Problems in BASIC (No. 1564—$15.50 paper; $19.95 hard)

Learning Simulation Techniques on a Microcomputer Playing Blackjack and Other Monte Carlo Games (No. 1535—$10.95 paper; $16.95 hard)

Making Money With Your Microcomputer (No. 1506—$8.25 paper)

Investment Analysis With Your Microcomputer (No. 1479—$13.50 paper; $19.95 hard)

The Art of Computer Programming (No. 1455—$10.95 paper; $16.95 hard)

25 Exciting Computer Games in BASIC for All Ages (No. 1427—$12.95 paper; $21.95 hard)

30 Computer Programs for the Homeowner, in BASIC (No. 1380—$10.25 paper; $16.95 hard)

Playing the Stock & Bond Markets with Your Personal Computer (No. 1251—$10.25 paper; $16.95 hard)

TAB | TAB BOOKS Inc.

Blue Ridge Summit. Pa. 17214

Send for FREE TAB Catalog describing over 750 current titles in print.